Regional Conference Series in Probability and Statistics
Volume 1

GROUP INVARIANCE APPLICATIONS IN STATISTICS

Morris L. Eaton
University of Minnesota

Institute of Mathematical Statistics
Hayward, California

American Statistical Association
Alexandria, Virginia

Conference Board of the Mathematical Sciences

Regional Conference Series
in Probability and Statistics

Supported by the
National Science Foundation

The production of the *Regional Conference Series in Probability and Statistics* is managed by the Institute of Mathematical Statistics: Paul Shaman, IMS Managing Editor; Jessica Utts, IMS Treasurer; and Jose L. Gonzalez, IMS Business Manager.

Library of Congress Catalog Card Number: 88-82946

International Standard Book Number 0-940600-15-3

Printed in the United States of America

Contents

Preface

These lecture notes are a result of the NSF/CBMS Regional Conference held at the University of Michigan, 15–19 June 1987. Topics in invariance with applications in statistics were discussed in a series of eighteen 45 minute lectures. Because of space and time limitations no material beyond that given in the lectures is presented here. Unfortunately, my original intention of including material on the invariance-sufficiency connection and on amenability had to be scrapped.

Professor Robb Muirhead who organized this conference is to be congratulated on a job well done. The keen interest and helpful suggestions of Professor Roger Farrell are here acknowledged. My thanks also go to Meera Somasundaram for turning my scribblings into a finished manuscript.

Morris L. Eaton

Minneapolis, March 1988

CHAPTER 1

Integrals and the Haar Measure

In this and the next chapter, background material on integration, topological groups and group actions is presented. The concrete examples described below provide a direct connection between the rather abstract theory of Haar measure and its application to situations which are relevant in statistical applications.

1.1. Integrals and measures. Our purpose in this section is to describe the connection between two different approaches to integration theory. The measure theory approach is commonly used in probability theory and so is familiar to most of us. However, the algebraic or linear functional approach is not so familiar. A good reference for the details of both approaches is Segal and Kunze (1978). A brief description of the linear functional approach and its relationship to the measure theory approach follows.

Let $\mathbf{X}$ be a locally compact topological space (Hausdorff) for which the topology has a countable base. That is, $\mathbf{X}$ is a topological space such that:

(i) The topology is Hausdorff.
(ii) Each point of $\mathbf{X}$ has a compact neighborhood.
(iii) The topology has a countable base.

Throughout these notes, the word "space" refers to a topological space $\mathbf{X}$ *satisfying* (i), (ii) *and* (iii). In virtually all of our examples, $\mathbf{X}$ is an open or closed subset of a finite dimensional vector space (with the usual Euclidean topology) so that $\mathbf{X}$ with its inherited topology is a space.

In many discussions of integration on locally compact spaces, (iii) is not assumed. However, when (iii) does hold the σ-algebra generated by the open sets (the Borel σ-algebra) and the σ-algebra generated by the compact sets (the Baire σ-algebra) are identical. This circumstance simplifies the discussion somewhat and from our point of view is not a serious restriction of generality. The σ-algebra associated with the space $\mathbf{X}$ is the Borel σ-algebra $\equiv$ Baire σ-algebra.

Given a space $\mathbf{X}$, let $K(\mathbf{X})$ be the set of all continuous real valued functions f defined on $\mathbf{X}$ which have compact support. Thus if $f \in K(\mathbf{X})$, there is a compact set $C \subset \mathbf{X}$ (which may vary from one f to another) such that $f(x) = 0$ if $x \notin C$. For $f_1, f_2 \in K(\mathbf{X})$ and $a_1, a_2 \in R$, it is clear that $a_1 f_1 + a_2 f_2 \in K(\mathbf{X})$ so that $K(\mathbf{X})$ is a real vector space.

DEFINITION 1.1. A real valued function J defined on $K(\mathbf{X})$ which satisfies

(i) $J(a_1 f_1 + a_2 f_2) = a_1 J(f_1) + a_2 J(f_2)$ for $f_1, f_2 \in K(\mathbf{X})$ and $a_1, a_2 \in R$,
(ii) $J(f) \geq 0$ whenever $f \in K(\mathbf{X})$ and f is nonnegative,
(iii) $J(f) > 0$ for some $f \in K(\mathbf{X})$

is called an *integral.*

An integral J is a linear function defined on the vector space $K(\mathbf{X})$ which maps nonnegative functions into nonnegative numbers. Assumption (iii) is to rule out the trivial case $J = 0$. Examples of integrals are easily constructed using Lebesgue measure when the space $\mathbf{X}$ is a subset of n dimensional Euclidean space R^n.

EXAMPLE 1.1. Assume $\mathbf{X} \subseteq R^n$ with its inherited topology is a space and let dx denote Lebesgue measure restricted to $\mathbf{X}$. If

$$\int_{\mathbf{X}} dx > 0,$$

define J on $K(\mathbf{X})$ by

$$J(f) = \int_{\mathbf{X}} f(x)\, dx.$$

Because the Lebesgue measure of compact sets is finite, $J(f)$ is a well defined finite number for $f \in K(\mathbf{X})$. That J is an integral follows easily. However, care must be taken in some cases. For example, take $\mathbf{X} = (-1, 1)$ and define a measure μ on the Borel sets of $\mathbf{X}$ using the function

$$g_0(x) = \begin{cases} \dfrac{1}{|x|}, & x \neq 0, \\ 0, & x = 0 \end{cases}$$

via the equation

$$\mu(B) = \int_B g_0(x)\, dx.$$

Then μ is a well defined σ-finite measure on $\mathbf{X}$ but μ does not define an integral via

$$f \to \int f(x) \mu(dx)$$

because there are functions in $K(\mathbf{X})$ for which

$$\int f(x) \mu(dx) = +\infty.$$

Of course, the problem arises because μ assigns measure $+\infty$ to some of the compact sets in $\mathbf{X}$. □

In order to describe the relationship between integrals (as they are defined above) and measures, Example 1.1 shows the same sort of restriction has to be made. Here is the appropriate definition which facilitates the identification of integrals with measures and vice versa.

DEFINITION 1.2. On a space $\mathbf{X}$, a measure μ defined on the Borel subsets of $\mathbf{X}$ is a *Radon measure* if $\mu(C) < +\infty$ for all compact sets $C \subseteq \mathbf{X}$ and $\mu \not\equiv 0$.

THEOREM 1.1. *Given an integral J defined on $K(\mathbf{X})$, there is a unique Radon measure μ such that*

$$J(f) = \int f(x)\mu(dx), \qquad f \in K(\mathbf{X}). \tag{1.1}$$

Conversely, a Radon measure $\mu \not\equiv 0$ defines an integral via (1.1).

The representation (1.1) sets up a one-to-one correspondence between integrals and Radon measures. In these notes, both representations are used with the choice being dictated by notational convenience and taste. Equation (1.1) allows the definition of an integral J to be extended to the class of functions which are μ-integrable. That is, if f is Borel-measurable and if

$$\int |f(x)|\mu(dx) < +\infty,$$

then the right-hand side of (1.1) is well defined and so $J(f)$ is defined via (1.1). In what follows, all integrals are automatically extended to the class of Borel-measurable and μ-integrable functions.

1.2. Topological groups. A group is a nonempty set G together with a binary operation $\circ$ such that the following conditions hold:

(i) $g_1, g_2 \in G$ implies $g_1 \circ g_2 \in G$.
(ii) $(g_1 \circ g_2) \circ g_3 = g_1 \circ (g_2 \circ g_3)$ for $g_1, g_2, g_3 \in G$.
(iii) There exists an element $e \in G$ such that $e \circ g = g \circ e = g$ for $g \in G$.
(iv) For each $g \in G$, there exists a unique element $g^{-1} \in G$ such that $g \circ g^{-1} = g^{-1} \circ g = e$.

The element $e \in G$ is called the *identity* and g^{-1} is called the *inverse* of g. It is customary to suppress the symbol $\circ$ when composing group elements. In this case, (ii) would be written $(g_1g_2)g_3 = g_1(g_2g_3)$. This custom is observed in what follows.

Now, assume that G is a group and that G is also a space as defined in Section 1.1. When the group operations fit together with the topology (i.e., are continuous), then G is a topological group. Here is the formal definition.

DEFINITION 1.3. Let the group G be a locally compact topological space (Hausdorff) whose topology has a countable base (i.e., G is a space). Assume that the functions

(i) $$(g_1, g_2) \to g_1 g_2$$

defined on $G \times G$ to G and

(ii) $$g \to g^{-1}$$

defined on G to G are both continuous. Then G is a *topological group*.

Again, the assumption that the topology for G has a countable base is one of convenience and is ordinarily not part of the definition of a topological group. In almost all the examples considered here, G is an open or closed subset of a finite dimensional vector space with the inherited topology. In the examples below, the verification that G is a topological group is not too hard and most of the details are left to the reader. Recall that $H \subset G$ is a *subgroup* of G if H, with its inherited operations from G, is a group.

EXAMPLE 1.2. Take $G = R^n$ with addition as the group operation and the usual topology on R^n. Obviously, $0 \in R^n$ is the identity and $-x$ is the inverse of $x \in R^n$. This group is *commutative*, that is, $x + y = y + x$, for $x, y \in R^n$. This example clearly extends to the case where G is a finite dimensional real vector space with the Euclidean topology. □

EXAMPLE 1.3. Let $G = \mathrm{Gl}_n$, which is the group of all $n \times n$ nonsingular real matrices with matrix multiplication as the group operation—commonly called the *general linear group*. For $g \in \mathrm{Gl}_n$, g^{-1} means the matrix inverse of g so that the identity in Gl_n is the $n \times n$ identity matrix. Let $\mathscr{L}_{n,n}$ be the real vector space of $n \times n$ real matrices so $\mathrm{Gl}_n \subseteq \mathscr{L}_{n,n}$. The determinant function det is defined on $\mathscr{L}_{n,n}$ and is continuous. Since

$$\mathrm{Gl}_n = \{x | x \in \mathscr{L}_{n,n}, \det(x) \neq 0\},$$

we see that Gl_n is an open subset of the Euclidean space $\mathscr{L}_{n,n}$. That is, Gl_n is the complement of the closed set $\{x | x \in \mathscr{L}_{n,n}, \det(x) = 0\}$. Thus, Gl_n is a space with the topology inherited from $\mathscr{L}_{n,n}$. That the group operations are continuous is not too hard to check. □

EXAMPLE 1.4. Let G be the group G_T^+ of $n \times n$ lower triangular matrices whose diagonal elements are positive. The group operation is matrix multiplication and G_T^+ is easily shown to be closed under this operation. That $g^{-1} \in G_T^+$ for $g \in G_T^+$ is most easily established by induction. For $g \in G_T^+$, partition g as

$$g = \begin{pmatrix} g_{11} & 0 \\ g_{21} & g_{22} \end{pmatrix},$$

where g_{11} is $(n-1) \times (n-1)$ and is lower triangular with positive diagonals,

g_{21} is $1 \times (n-1)$ and $g_{22} \in (0, \infty)$. The inverse of g is

$$\begin{pmatrix} g_{11}^{-1} & 0 \\ -g_{22}^{-1}g_{21}g_{11}^{-1} & g_{22}^{-1} \end{pmatrix},$$

which shows that $g^{-1} \in G_T^+$. The topology for G_T^+ is that obtained by regarding G_T^+ as a subset of $n(n+1)/2$ dimensional coordinate space and using the inherited topology for G_T^+. Clearly G_T^+ is an open subset of this $n(n+1)/2$ dimensional vector space. The continuity of the group operations follows from the fact that G_T^+ is a (closed) subgroup of Gl_n and hence inherits the continuity of the group operations from Gl_n. □

EXAMPLE 1.5. Let G be the group O_n of $n \times n$ real orthogonal matrices. Since O_n is a closed bounded set in $\mathscr{L}_{n,n}$, O_n is a compact space with its inherited topology. Further, O_n is a closed subgroup of Gl_n and its topology is equal to that which it inherits from Gl_n. Thus the group operations in O_n are continuous. O_n is called the *orthogonal group*. □

EXAMPLE 1.6. In this example, we describe what is usually called the *affine group*, Al_n. Elements of Al_n are pairs (g, x) with $g \in \mathrm{Gl}_n$ and $x \in R^n$. The group operation is defined by

$$(g_1, x_1)(g_2, x_2) = (g_1 g_2, g_1 x_2 + x_1),$$

where $g_1 g_2$ is the composition of g_1 and g_2 in Gl_n and $g_1 x_2 + x_1$ means the matrix g_1 multiplied times x_2 with x_1 added to $g_1 x_2$. The identity in Al_n is $(e, 0)$ where e is the identity in Gl_n so

$$(g, x)^{-1} = (g^{-1}, -g^{-1}x).$$

The topology for Al_n is that of the product space $\mathrm{Gl}_n \times R^n$, which is an open subset of the coordinate space $\mathscr{L}_{n,n} \times R^n$. The continuity of the group operations is left to the reader to check. □

Finally, we mention two discrete groups which arise in statistical applications. The first of these is $\mathbf{P}_n$, the group of *permutation matrices*. Elements of $\mathbf{P}_n$ are $n \times n$ matrices g such that in each row and each column of g, there is exactly one element which is equal to 1 and all the remaining elements are 0. Each $g \in \mathbf{P}_n$ is an orthogonal matrix because gx and x have the same length for all $x \in R^n$. Thus $g^{-1} = g'$ where g' is the transpose of g. That $\mathbf{P}_n$ is closed under matrix multiplication is easily checked. Of course, $\mathbf{P}_n$ is a topological group with the discrete topology. Obviously $\mathbf{P}_n$ has $n!$ elements. Elements of $\mathbf{P}_n$ are called permutation matrices because for each $x \in R^n$, gx is a vector whose coordinates are a permutation of the coordinates of x.

The group of *coordinate sign changes* $\mathbf{D}_n$ has elements g which are $n \times n$ matrices with diagonal elements equal to ± 1 and off diagonal elements 0. Obviously each g is an orthogonal matrix and $\mathbf{D}_n$ has 2^n elements.

1.3. Haar measure. One of the most important theorems regarding topological groups asserts the existence and uniqueness (up to a positive constant) of left- and right-invariant integrals (measures). This theorem together with the relationship between left- and right-invariant integrals is described here. A more extensive discussion of these results together with proofs can be found in Nachbin (1965) and Segal and Kunze (1978).

To state things precisely let G be a topological group as in Definition 1.3. The real vector space $K(G)$ is the set of all continuous functions with compact support defined on G. Given $g \in G$, define the transformation L_g on $K(G)$ to $K(G)$ by

$$(L_g f)(x) = f(g^{-1}x)$$

for $g, x \in G$ and $f \in K(G)$.

DEFINITION 1.4. An integral J on $K(G)$ is *left-invariant* if for all $f \in K(G)$,

$$J(L_g f) = J(f) \quad \text{for } g \in G.$$

If μ is the Radon measure corresponding to a left-invariant integral, then μ satisfies

$$\int_G f(g^{-1}x)\mu(dx) = \int_G f(x)\mu(dx), \qquad g \in G,$$

for $f \in K(G)$ and hence for all μ-integrable f. Alternatively, this condition is sometimes written less formally as

$$\mu(d(gx)) = \mu(dx), \qquad g \in G,$$

by simply defining $\mu(d(gx))$ as

$$\int f(x)\mu(d(gx)) = \int f(g^{-1}x)\mu(dx).$$

Here is the basic theorem.

THEOREM 1.2. *On a topological group G, there exists a left-invariant integral (measure). If J_1 and J_2 are left-invariant integrals, then there exists a positive constant c such that $J_1 = cJ_2$.*

In these notes, we move freely between integrals and the corresponding Radon measure. If J is a left-invariant integral on $K(G)$, the corresponding measure, denoted by ν_l, satisfies

$$J(f) = \int_G f(x)\nu_l(dx) = \int_G f(g^{-1}x)\nu_l(dx)$$

for $g \in G$. The measure ν_l is often called a *left Haar measure*—named after A. Haar who first established the existence of invariant measures on many topological groups. Since an invariant integral is unique up to a positive constant, so is ν_l.

To describe the relationship between left- and right-invariant integrals, it is convenient to introduce the so-called (right-hand) modulus of G. Let ν_l be a left Haar measure and define J_1 on $K(G)$ by

$$J_1(f) = \int f(xg^{-1})\nu_l(dx)$$

for a fixed $g \in G$. It is easily verified that J_1 is left-invariant as is

$$J(f) = \int f(x)\nu_l(dx).$$

By the uniqueness portion of Theorem 1.2, there is a positive constant which is denoted by $\Delta(g)$ such that $J_1 = \Delta(g)J$. Thus, Δ is defined by

$$\int f(xg^{-1})\nu_l(dx) = \Delta(g)\int f(x)\nu_l(dx), \qquad f \in K(G). \tag{1.2}$$

THEOREM 1.3. *The function Δ defined on G to $(0, \infty)$, called the modulus of G, is continuous and satisfies*

$$\Delta(g_1g_2) = \Delta(g_1)\Delta(g_2).$$

In particular $\Delta(e) = 1$ *and* $\Delta(g^{-1}) = 1/\Delta(g)$.

PROOF. A proof of the continuity can be found in Nachbin [(1965), page 77]. To verify the functional equation for Δ, compute

$$\begin{aligned}\Delta(g_1g_2)\int f(x)\nu_l(dx) &= \int f\big(x(g_1g_2)^{-1}\big)\nu_l(dx)\\ &= \int f\big((xg_2^{-1})g_1^{-1}\big)\nu_l(dx)\\ &= \Delta(g_1)\int f\big(xg_2^{-1}\big)\nu_l(dx)\\ &= \Delta(g_1)\Delta(g_2)\int f(x)\nu_l(dx).\end{aligned}$$ □

To define right-invariant integrals, for $g \in G$ define R_g on $K(G)$ by

$$(fR_g)(x) = f(xg^{-1}).$$

Note that R_g is written on the right of f so that the equation

$$R_{g_1}R_{g_2} = R_{g_1g_2}$$

is valid. (More is said about this when group actions are discussed.)

DEFINITION 1.5. An integral J is *right-invariant* if $J(f) = J(fR_g)$ for $f \in K(G)$ and $g \in G$.

THEOREM 1.4. *Let J be a left-invariant integral given by*

$$J(f) = \int f(x)\nu_l(dx)$$

and let Δ *be the modulus of* G. *Then the integral* J_1 *defined by*

$$J_1(f) = \int f(x)\Delta(x^{-1})\nu_l(dx)$$

is right-invariant and satisfies

$$\int f(x)\Delta(x^{-1})\nu_l(dx) = \int f(x^{-1})\nu_l(dx) \tag{1.3}$$

for $f \in K(G)$.

PROOF. That J_1 is right-invariant follows from the definition of Δ and the calculation

$$\begin{aligned} J_1(fR_g) &= \int f(xg^{-1})\Delta(x^{-1})\nu_l(dx) \\ &= \int f(xg^{-1})\frac{1}{\Delta(xg^{-1}g)}\nu_l(dx) \\ &= \frac{1}{\Delta(g)}\int \frac{f(xg^{-1})}{\Delta(xg^{-1})}\nu_l(dx) \\ &= \frac{\Delta(g)}{\Delta(g)}\int \frac{f(x)}{\Delta(x)}\nu_l(dx) = J_1(f). \end{aligned}$$

A proof of the second equation can be found in Nachbin [(1965), page 78]. □

Equation (1.3) establishes a relationship between left and right Haar measure. Namely, if ν_l is a left Haar measure, then

$$\nu_r(dx) = \Delta(x^{-1})\nu_l(dx) \tag{1.4}$$

is a right Haar measure. Further if ν_r is a right Haar measure, a similar construction shows that

$$\nu_l(dx) = \Delta(x)\nu_r(dx) \tag{1.5}$$

is a left Haar measure.

Before turning to a few examples, we discuss the important special case of compact groups. Here is a characterization of compact groups in terms of Haar measure.

THEOREM 1.5. *In order that G be compact, it is necessary and sufficient that there exist a finite Haar measure* [*that is,* $\mu(G) < +\infty$].

PROOF. See Nachbin [(1965), page 75]. □

Let Δ be the modulus of a compact group G. We claim that $\Delta \equiv 1$. To see this, recall that Δ is a continuous function from G into $(0, \infty)$ which satisfies

$$\Delta(g_1 g_2) = \Delta(g_1)\Delta(g_2),$$

that is, Δ is a group homomorphism. Thus $H = \{\Delta(g) | g \in G\}$ is a subgroup of the multiplicative group $(0, \infty)$. However, H is compact as it is the continuous image of the compact set G. If $H \neq \{1\}$, then there exists $a \in H$ such that $a > 1$. Hence $a^n \in H$ for $n = 1, 2, \ldots$. But $\{a^n | n = 1, 2, \ldots\}$ does not contain a convergent subsequence which contradicts the compactness of H. Thus $\Delta \equiv 1$ for any compact group G. Therefore all right Haar measures are left Haar measures and conversely. In these notes, the Haar measure on a compact group G is always normalized so that $\mu(G) = 1$, that is, the Haar measure is taken to be a probability measure.

Here are some examples of left and right Haar measures.

EXAMPLE 1.7. Take G to be the vector space R^n with addition as the group operation. Then ordinary Lebesgue measure on R^n is both left- and right-invariant (the group here is commutative) and so the modulus is $\Delta = 1$. □

EXAMPLE 1.8. If G is any finite or denumerable discrete group, then counting measure is both left- and right-invariant and so the modulus is 1. □

EXAMPLE 1.9. For this example, take $G = \mathrm{Gl}_n$ as in Example 1.3. Let dx denote Lebesgue measure restricted to Gl_n with Gl_n regarded as a subset of $\mathscr{L}_{n,n}$. Since G is open in $\mathscr{L}_{n,n}$, Gl_n has positive Lebesgue measure. Define an integral J by

$$J(f) = \int f(x) \frac{dx}{|\det(x)|^n}.$$

The claim is that J is both left- and right-invariant. To see this, first consider

$$J(L_g f) = \int f(g^{-1}x) \frac{dx}{|\det(x)|^n}$$

and make the change of variables $y = g^{-1}x$ so $x = gy$. To calculate the Jacobian of this transformation (defined by g), write x as a column vector

$$\begin{pmatrix} x^1 \\ x^2 \\ \vdots \\ x^n \end{pmatrix} \in R^{n^2},$$

where x^i is the ith column of x and do the same with y. In this notation, the equation $x = gy$ becomes

$$\begin{pmatrix} x^1 \\ x^2 \\ \vdots \\ x^n \end{pmatrix} = \begin{pmatrix} g & & & \\ & g & & \\ & & \ddots & \\ & & & g \end{pmatrix} \begin{pmatrix} y^1 \\ y^2 \\ \vdots \\ y^n \end{pmatrix},$$

where the elements not indicated in the $n^2 \times n^2$ matrix are 0. Obviously this linear transformation relating x to y has determinant equal to $(\det(g))^n$. Thus $x = gy$ implies

$$dx = |(\det(g))^n|\, dy = |\det(g)|^n\, dy.$$

Substituting this into the expression for $J(L_g f)$ yields

$$J(L_g f) = \int f(g^{-1}x) \frac{dx}{|\det(x)|^n} = \int f(y) \frac{|\det(g)|^n\, dy}{|\det(gy)|^n}$$

$$= \int f(y) \frac{dy}{|\det(y)|^n} = J(f).$$

Thus J is a left-invariant integral. A similar calculation (using rows of x and y instead of columns) shows that J is also right-invariant. Thus, the modulus of Gl_n is identically 1. □

Groups for which the modulus is 1 are called *unimodular*. Clearly all commutative groups are unimodular as are finite or denumerable discrete groups and compact groups. In contrast to Gl_n, the following example shows that Al_n is not unimodular.

EXAMPLE 1.10. Take $G = \mathrm{Al}_n$ so elements of Al_n are (x, a) with $x \in \mathrm{Gl}_n$ and $a \in R^n$. That Al_n can be regarded as an open subset of $\mathscr{L}_{n,n} \times R^n$ was discussed in Example 1.6. Let $dx\, da$ denote Lebesgue measure on $\mathscr{L}_{n,n} \times R^n$ restricted to Al_n. Define an integral J by

$$J(f) = \int f[(x, a)] \frac{dx\, da}{|\det(x)|^{n+1}}.$$

We now show that J is left-invariant. With $g \in \mathrm{Gl}_n$ and $\alpha \in R^n$, $h = (g, \alpha) \in \mathrm{Al}^n$ and

$$J(L_h f) = \int f[h^{-1}(x, a)] \frac{dx\, da}{|\det(x)|^{n+1}}.$$

Set $(y, b) = h^{-1}(x, a)$ so that

$$(x, a) = h(y, b) = (gy, gb + \alpha).$$

Thus $x = gy$ and $a = gb + \alpha$. From the last example, the Jacobian of the transformation $x = gy$ is $|\det(g)|^n$ and it is obvious that the Jacobian of the transformation $a = gb + \alpha$ is $|\det(g)|$. Making this change of variables, we have

$$J(L_h f) = \int f[(y, b)] \frac{|\det(g)|^{n+1}}{|\det(gy)|^{n+1}}\, dy\, db$$

$$= \int f[(y, b)] \frac{dy\, db}{|\det(y)|^{n+1}} = J(f).$$

A similar (but slightly different) calculation shows that

$$J_1(f) = \int f[(x, a)] \frac{dx\, da}{|\det(x)|^n}$$

is a right-invariant integral. From Equation (1.4), it follows that the modulus of Al_n is

$$\Delta(x, a) = \frac{1}{|\det(x)|}. \qquad \square$$

Other examples of invariant integrals are given in the next section after the discussion of multipliers and relatively invariant integrals.

1.4. Multipliers and relatively invariant integrals. The modulus Δ of a topological group G provides an example of a multiplier, namely a continuous homomorphism of G into the multiplicative group of $(0, \infty)$. Multipliers play the role of Jacobians for some integrals on $K(G)$.

DEFINITION 1.6. A function χ defined on G to $(0, \infty)$ is a *multiplier* if χ is continuous and $\chi(g_1 g_2) = \chi(g_1)\chi(g_2)$ for $g_1, g_2 \in G$.

It is clear that $\chi(e) = 1$ and $\chi(g_1 g_2) = \chi(g_2 g_1)$ for any multiplier χ, so that $\chi(g^{-1}) = 1/\chi(g)$. Further, $\chi(G) = \{\chi(g) | g \in G\}$ is a subgroup of $(0, \infty)$. Thus, if G is compact, $\chi(G)$ must be a compact subgroup of $(0, \infty)$ and thus $\chi(G) = \{1\}$. Hence the trivial multiplier, $\chi = 1$, is the only multiplier defined on a compact group.

DEFINITION 1.7. An integral J on $K(G)$ is *relatively* (left) *invariant* with multiplier χ if for each $g \in G_1$,

$$J(L_g f) = \int f(g^{-1}x)m(dx) = \chi(g)\int f(x)m(dx) = \chi(g)J(f).$$

The relationship between invariant and relatively invariant integrals is given in the following result.

THEOREM 1.6. *Let χ be a multiplier on G.*

(i) *If $J_1(f) = \int f(x)m(dx)$ is relatively invariant with multiplier χ, then*

$$J(f) = \int f(x)\chi(x^{-1})m(dx)$$

is left-invariant.

(ii) *If $J(f) = \int f(x)\nu_l(dx)$ is left-invariant, then*

$$J_1(f) = \int f(x)\chi(x)\nu_l(dx)$$

is relatively invariant with multiplier χ.

PROOF. Only the proof of (i) is given, as the proof of (ii) is similar. For $g \in G$,

$$J(L_g f) = \int f(g^{-1}x)\chi(x^{-1})m(dx) = \int f(g^{-1}x)\chi\big((g^{-1}x)^{-1}g^{-1}\big)m(dx)$$

$$= \chi(g^{-1})\chi(g)\int f(x)\chi(x^{-1})m(dx) = J(f).$$

Thus, J is left invariant. □

Theorem 1.6 shows how to construct a relatively invariant integral with a given multiplier from a left-invariant measure, and conversely. In practice, it is the converse which is somewhat more useful since it provides a very useful method of constructing a left-invariant integral in some cases. Here is a rather informal description of the method which is used below to construct a left-invariant measure on the group G_T^+ of Example 1.4. Suppose the topological group G is subset of a Euclidean space R^m such that G has positive Lebesgue measure. Define a integral on $K(G)$ by

$$J_1(f) = \int_G f(x)\,dx,$$

where dx denotes Lebesgue measure restricted to G. If we can show that

$$\int_G f(g^{-1}x)\,dx = \chi(g)\int_G f(x)\,dx \quad \text{for } g \in G, \tag{1.6}$$

then

$$J(f) = \int_G f(x)\chi(x^{-1})\,dx$$

is left-invariant. The verification of (1.6) is ordinarily carried out by changing variables and computing a Jacobian.

One can also define relatively right-invariant integrals, but their relationship to relatively left integrals follows from (1.5) and the equation

$$m(dx) = \chi(x)\nu_l(dx), \tag{1.7}$$

which relates the left Haar measure ν_l and the measure m which defines a relatively invariant integral with multiplier χ. For this reason, relatively right-invariant integrals are not mentioned explicitly.

It follows from Theorem 1.6 that the problem of describing all the relatively invariant integrals on $K(G)$ can be solved by first exhibiting one relatively invariant integral and then describing all the multipliers on G. This program is carried out in some examples that follow. Before turning to these examples, it is useful to describe a few general facts concerning multipliers.

THEOREM 1.7. *Suppose the topological group G is the direct product $G_1 \times G_2$ of two topological groups [that is, each $g \in G$ has the form $g = (g_1, g_2)$ with $g_i \in G_i$ and the group operation is $g \circ h = (g_1h_1, g_2h_2)$ where $g = (g_1, g_2)$ and*

$h = (h_1, h_2)$]. *The topology on G is assumed to be the product topology. Then* χ *is a multiplier on G iff*

$$\chi[(g_1, g_2)] = \chi_1(g_1)\chi_2(g_2), \tag{1.8}$$

where χ_i *is a multiplier on* G_i, $i = 1, 2$.

PROOF. If χ has the form (1.8), it is clear that χ is a multiplier. Conversely, suppose that χ is a multiplier. For $g = (g_1, g_2) \in G$, write

$$(g_1, g_2) = (g_1, e_2)(e_1, g_2),$$

where e_i is the identity in G_i, $i = 1, 2$. Then consider

$$\chi_1(g_1) = \chi[(g_1, e_2)], \qquad g_1 \in G_1,$$
$$\chi_2(g_2) = \chi[(e_1, g_2)], \qquad g_2 \in G_2.$$

The continuity of χ_i on G_i follows since χ is continuous and χ_i is χ restricted to a closed subset of G. That (1.8) holds follows since χ is a homomorphism. □

Also observe that if χ is a multiplier on G and H is a closed subgroup of G, then the restriction of χ to H is a multiplier on H. In particular, if H is a compact subgroup of G, then $\chi(h) = 1$ for all $h \in H$ since compact groups have only trivial multipliers.

EXAMPLE 1.11. In this example, all the relatively invariant integrals on R^n are computed. Because R^n is an n-fold product group, Theorem 1.7 shows that it is sufficient to treat the case $n = 1$. Thus, suppose that χ_1 is a continuous function from R^1 to $(0, \infty)$ which satisfies the equation

$$\chi_1(x + y) = \chi_1(x)\chi_1(y).$$

Taking logs of both sides of this leads to

$$\log \chi_1(x + y) = f(x + y) = f(x) + f(y)$$

for $x, y \in R^1$. This is the well known Cauchy functional equation which has the solution (because f is continuous)

$$f(x) = cx,$$

where c is some real number. Thus,

$$\chi_1(x) = \exp[cx], \qquad x \in R^1,$$

where c is a fixed real number. Applying Theorem 1.7 shows that χ is a multiplier on R^n iff χ has the form

$$\chi(u) = \exp\left[\sum_1^n c_i u_i\right],$$

where $u \in R^n$ has coordinates $u_1, \ldots, u_n$ and $c_1, \ldots, c_n$ are fixed real numbers. Thus, a Radon measure m is relatively invariant with multiplier χ given above iff

$$m(du) = c_0 \chi(u)\, du, \qquad u \in R^n,$$

where c_0 is a fixed constant and du is Lebesgue measure on R^n.

When G is the product multiplicative group $(0, \infty)^n$, the continuous isomorphism ϕ: $G \to R^n$ given by

$$\phi(x) = \begin{pmatrix} \log x_1 \\ \log x_2 \\ \vdots \\ \log x_n \end{pmatrix} \in R^n, \qquad x \in (0, \infty)^n,$$

can be used to show that the measure

$$\nu_l(dx) = \frac{dx_1 \cdots dx_n}{\prod_{i=1}^n x_i}$$

is right- and left-invariant on G. Further, χ is a multiplier on G iff χ has the form

$$\chi(x) = \prod_{i=1}^n (x_i)^{c_i},$$

where $c_1, \dots, c_n$ are fixed real numbers. Thus, all the relatively invariant integrals on $K(G)$ are known. □

EXAMPLE 1.12. Consider $G = \mathrm{Gl}_n$ as in Example 1.3 so

$$\nu_l(dx) = \frac{dx}{|\det(x)|^n}$$

is left- and right-invariant. To describe all the multipliers on Gl_n, the functional equation

(1.9) $$\chi(xy) = \chi(x)\chi(y), \qquad x, y \in \mathrm{Gl}_n,$$

must be solved. To this end, recall the singular value decomposition of an $n \times n$ matrix $x \in \mathrm{Gl}_n$; it is

$$x = \alpha D \beta,$$

where $\alpha, \beta \in O_n$ and D is an $n \times n$ diagonal matrix with positive diagonal elements [for example, see Eaton (1983), page 58]. Thus if χ is continuous and satisfies (1.9), then

$$\chi(x) = \chi(\alpha D \beta) = \chi(\alpha)\chi(D)\chi(\beta) = \chi(D)$$

since α and β are both elements of the compact group O_n so $\chi(\alpha) = \chi(\beta) = 1$. Now, write D as

$$D = \prod_{i=1}^n D_i(d_i),$$

where d_i is the ith diagonal of D and $D_i(d_i)$ has ith diagonal equal to d_i and all other diagonal elements equal to 1. Then, because $D_i(d_i) \in \mathrm{Gl}_n$,

$$\chi(D) = \chi\left(\prod_{i=1}^n D_i(d_i)\right) = \prod_{i=1}^n \chi(D_i(d_i)).$$

A bit of reflection shows that there exists a permutation matrix $P \in \mathbf{P}_n \subset O_n$ such that

$$D_i(d_i) = PD_1(d_i)P', \qquad i = 1, \ldots, n,$$

so that

$$\chi(D_i(d_i)) = \chi(D_1(d_i)), \qquad i = 1, \ldots, n,$$

because $\chi(P) = \chi(P') = 1$. Thus,

$$\chi(D) = \prod_{i=1}^{n} \chi(D_i(d_i)) = \prod_{i=1}^{n} \chi(D_1(d_i))$$
$$= \chi\left(\prod_{i=1}^{n} D_1(d_i)\right) = \chi\left(D_1\left(\prod_{i=1}^{n} d_i\right)\right).$$

This shows that χ is determined by its values on matrices of the form

$$D_1(\lambda) = \begin{pmatrix} \lambda & & & & 0 \\ & 1 & & & \\ & & 1 & & \\ & & & \ddots & \\ 0 & & & & 1 \end{pmatrix} \in \mathrm{Gl}_n, \qquad \lambda > 0.$$

Since these matrices form a subgroup of Gl_n which is isomorphic to the multiplicative group $(0, \infty)$, it follows from Example 1.11 that

$$\chi(D_1(\lambda)) = \lambda^c$$

for some fixed real number c. Now, retracing the above steps, we have

$$\chi(x) = \chi(D) = \left(\prod_{i=1}^{n} d_i\right)^c$$
$$= |\det(x)|^c.$$

Hence, all multipliers on Gl_n are some power of $|\det(x)|$. The converse is clear. Thus every relatively invariant measure has the form

$$c_0 |\det(x)|^c \frac{dx}{|\det(x)|^n}$$

for some fixed number c and some $c_0 > 0$. □

EXAMPLE 1.13. With $G = \mathrm{Al}_n$ of Example 1.6, recall that the group operation in Al_n is

$$(g_1, x_1)(g_2, x_2) = (g_1 g_2, g_1 x_2 + x_1)$$

for $g_i \in \mathrm{Gl}_n$ and $x_i \in R_n$, $i = 1, 2$. To describe the multipliers on Al_n, consider χ on Al_n and write $(g, x) \in \mathrm{Al}_n$ as

$$(g, x) = (g, 0)(e, g^{-1}x) = (e, x)(g, 0),$$

where e is the identity in Gl_n. Thus, a multiplier χ must satisfy

$$\chi(g, x) = \chi(g, 0)\chi(e, g^{-1}x) = \chi(g, 0)\chi(e, x)$$

so

$$\chi(e, g^{-1}x) = \chi(e, x)$$

for all $g \in \mathrm{Gl}_n$. Letting g^{-1} converge to 0, the continuity of χ implies that

$$1 = \chi(e, 0) = \chi(e, x), \qquad x \in R^n.$$

Hence

$$\chi(g, x) = \chi(g, 0), \qquad g \in \mathrm{Gl}_n.$$

But $g \to \chi(g, 0)$ defines a multiplier on Gl_n so by the previous example,

$$\chi(g, x) = |\det(g)|^c$$

for some fixed real number c. Conversely, it is clear that such a function on Al_n is a multiplier. Thus, all the multipliers on Al_n have been given. □

EXAMPLE 1.14. The final example of this chapter concerns $G_T^+ \subset \mathrm{Gl}_n$. The left and right Haar measures, the modular function and all the multipliers for G_T^+ are derived here. We are going to use the method described after the proof of Theorem 1.6. For $x \in G_T^+$, x has the form

$$x = \begin{pmatrix} x_{11} & 0 & \cdots & 0 \\ x_{21} & x_{22} & \cdots & 0 \\ \vdots & \vdots & & \vdots \\ x_{n1} & x_{n2} & \cdots & x_{nn} \end{pmatrix}$$

with $x_{ii} > 0$ and $x_{ij} \in R^n$, $i > j$. Let dx denote Lebesgue measure restricted to the set of such x's in $n(n+1)/2$ dimensional space. Define an integral J on $K(G_T^+)$ by

$$J(f) = \int f(x)\, dx$$

and for $g \in G_T^+$, consider

$$J(L_g f) = \int f(g^{-1}x)\, dx.$$

With $y = g^{-1}x$ so $x = gy$, the Jacobian of this transformation on $n(n+1)/2$ coordinate space is

$$\chi_0(g) = \prod_{i=1}^{n} g_{ii}^i,$$

where $g_{11}, \ldots, g_{nn}$ are the diagonal elements of $g \in G_T^+$ [see Eaton (1983), page 171]. Hence $dx = \chi_0(g)\, dy$ so that

$$J(L_g f) = \chi_0(g) J(f).$$

From Theorem 1.6, we have that

$$\nu_l(dx) = \frac{dx}{\chi_0(x)} = \frac{dx}{\prod_{i=1}^n x_{ii}^i} \tag{1.10}$$

is a left Haar measure on $G_T^+ \subset \mathrm{Gl}_n$. The modulus of G_T^+ is computed using the definition (1.2). Thus, consider

$$\int f(xg^{-1})\nu_l(dx) = \int f(xg^{-1})\frac{dx}{\chi_0(x)}$$

and let $y = xg^{-1}$ so $x = yg$. The Jacobian of this transform is

$$\chi_1(g) = \prod_{i=1}^{n} g_{ii}^{n-i+1},$$

where $g_{11}, \ldots, g_{nn}$ are the diagonal elements of g. Making this change of variables in the above integral yields

$$\begin{aligned}\int f(xg^{-1})\nu_l(dx) &= \int f(y)\chi_1(g)\frac{dy}{\chi_0(yg)}\\ &= \frac{\chi_1(g)}{\chi_0(g)}\int f(y)\frac{dy}{\chi_0(y)}\\ &= \frac{\chi_1(g)}{\chi_0(g)}\int f(x)\nu_l(dx).\end{aligned}$$

Thus, by definition, the modulus of G_T^+ is

$$\Delta(g) = \frac{\chi_1(g)}{\chi_0(g)} = \prod_{i=1}^{n} g_{ii}^{n-2i+1} \tag{1.11}$$

so a right Haar measure is

$$\nu_r(dx) = \frac{1}{\Delta(x)}\nu_l(dx) = \frac{dx}{\prod_{i=1}^{n} x_{ii}^{n-i+1}}. \tag{1.12}$$ □

Here is a characterization of the multipliers on $G_T^+ \subset \mathrm{Gl}_n$.

Theorem 1.8. *All multipliers on $G_T^+ \subset \mathrm{Gl}_n$ have the form*

$$\chi(g) = \prod_{i=1}^{n} (g_{ii})^{c_i},$$

where $c_1, \ldots, c_n$ are fixed real numbers. Conversely, any such function is a multiplier.

Proof. That such a function is a multiplier is easily checked. To show that all multipliers have the claimed form, we argue by induction. For $n = 1$, $G_T^+ = (0, \infty)$ and the assertion follows from Example 1.11. Assume the result is true for $G_T^+ \subset \mathrm{Gl}_n$ and consider $G_T^+ \subset \mathrm{Gl}_{n+1}$. For $g \in G_T^+ \subset \mathrm{Gl}_{n+1}$, partition g as

$$g = \begin{pmatrix} x_{11} & 0 \\ x_{21} & x_{22} \end{pmatrix},$$

where x_{11} is $n \times n$, x_{21} is $1 \times n$ and x_{22} is in $(0, \infty)$. With e denoting the $n \times n$

identity matrix, observe that

$$\begin{pmatrix} x_{11} & 0 \\ x_{21} & x_{22} \end{pmatrix} = \begin{pmatrix} x_{11} & 0 \\ 0 & 1 \end{pmatrix}\begin{pmatrix} e & 0 \\ x_{21} & x_{22} \end{pmatrix} = \begin{pmatrix} e & 0 \\ x_{21}x_{11}^{-1} & x_{22} \end{pmatrix}\begin{pmatrix} x_{11} & 0 \\ 0 & 1 \end{pmatrix}.$$

Thus, if χ is a multiplier,

$$\chi\begin{pmatrix} x_{11} & 0 \\ 0 & 1 \end{pmatrix}\chi\begin{pmatrix} e & 0 \\ x_{21} & x_{22} \end{pmatrix} = \chi\begin{pmatrix} e & 0 \\ x_{21}x_{11}^{-1} & x_{12} \end{pmatrix}\chi\begin{pmatrix} x_{11} & 0 \\ 0 & 1 \end{pmatrix}$$

so that

$$\chi\begin{pmatrix} e & 0 \\ x_{21} & x_{22} \end{pmatrix} = \chi\begin{pmatrix} e & 0 \\ x_{21}x_{11}^{-1} & x_{22} \end{pmatrix}.$$

Letting x_{11}^{-1} converge to 0, the continuity of χ implies that

$$\chi\begin{pmatrix} e & 0 \\ x_{21} & x_{22} \end{pmatrix} = \chi\begin{pmatrix} e & 0 \\ 0 & x_{22} \end{pmatrix}.$$

But, the function

$$x_{22} \to \chi\begin{pmatrix} e & 0 \\ 0 & x_{22} \end{pmatrix}$$

defines a multiplier on $(0, \infty)$, so

$$\chi\begin{pmatrix} e & 0 \\ 0 & x_{22} \end{pmatrix} = (x_{22})^{c_{n+1}}$$

for some real c_{n+1}. Also, the function

$$x_{11} \to \chi\begin{pmatrix} x_{11} & 0 \\ 0 & 1 \end{pmatrix}$$

defines a multiplier on $G_T^+ \subset \mathrm{Gl}_n$. Since the diagonal elements of x_{11} are $g_{11}, \ldots, g_{n,n}$ and $x_{22} = g_{n+1,n+1}$, using the induction hypothesis, we have

$$\begin{aligned}\chi(g) = \chi\begin{pmatrix} x_{11} & 0 \\ x_{21} & x_{22} \end{pmatrix} &= \chi\begin{pmatrix} x_{11} & 0 \\ 0 & 1 \end{pmatrix}\chi\begin{pmatrix} e & 0 \\ 0 & x_{22} \end{pmatrix} \\ &= \left(\prod_{i=1}^{n} g_{ii}^{c_i}\right) g_{n+1}^{c_{n+1}} = \prod_{i=1}^{n+1} g_{ii}^{c_i}.\end{aligned}$$

This completes the proof. □

That the compact group O_n has a unique left- and right-invariant probability measure follows from the general theory of Haar measure. However, the explicit construction of this probability, in terms of a "coordinate system" for O_n is a bit technical and requires some familiarity with differential forms. This topic is not discussed here, but in subsequent material, it is shown how to construct a random orthogonal matrix in O_n whose distribution is exactly the Haar measure on O_n. This construction involves the multivariate normal distribution.

CHAPTER 2

Group Actions and Relatively Invariant Integrals

In this chapter, group actions are reviewed and are illustrated with examples of relevance for statistical applications. Relatively invariant integrals (measures) are defined and examples are given. An important result, due to Weil, gives necessary and sufficient conditions for the existence and uniqueness of relatively invariant integrals when the group action is transitive. A discussion of invariant and equivariant functions closes out the chapter.

2.1. Group actions. In many examples, the elements of a group G are functions which are one-to-one and onto from a set $\mathbf{X}$ to itself. Further, the group operation in G is just function composition when the elements of G are regarded as functions on $\mathbf{X}$. A typical example of this is the group Gl_n of $n \times n$ nonsingular matrices and the space R^n. Each $g \in \mathrm{Gl}_n$ is a one-to-one onto map from R^n to R^n and

$$g_1(g_2(x)) = (g_1 \circ g_2)(x),$$
$$ex = x.$$

That is, g_1 evaluated at the point $g_2(x) \in R^n$ is equal to $g_1 \circ g_2 \in \mathrm{Gl}_n$ evaluated at $x \in R^n$ and the identity in G is the identity function. Thus, function composition is "the same as" the group operation. In many circumstances the group operation is defined so that the above relationship holds. The idea of a group action on a set simply abstracts the essentials of this situation.

Let $\mathbf{X}$ be a set and let G be a group with identity e.

DEFINITION 2.1. A function F defined on $G \times \mathbf{X}$ to $\mathbf{X}$ satisfying

(i) $F(e, x) = x$, $x \in \mathbf{X}$,
(ii) $F(g_1 g_2, x) = F(g_1, F(g_2, x))$, $g_1, g_2 \in G$, $x \in \mathbf{X}$,

specifies G *acting on the left of* $\mathbf{X}$.

Although Definition 2.1 captures mathematically what one means by a left group action, in concrete examples, the explicit use of F can be mathematical overkill. In the statistical literature, the most common verbiage to indicate left group action is "Suppose G acts on the left of $\mathbf{X}$ with action $x \to gx$." What this means is that the value of F at (g, x) is denoted by gx (whose definition is suppose to be clear from context), so in this notation, conditions (i) and (ii) of Definition 2.1 are

$$ex = x, \qquad x \in \mathbf{X},$$
$$(g_1 g_2)x = g_1(g_2 x).$$

That is, $g \in G$ is thought of as defining a function of $\mathbf{X}$ to $\mathbf{X}$ and the value of g at x is written gx or sometimes $g(x)$. The equation $(g_1 g_2)x = g_1(g_2 x)$ then means that "function composition" and the group operation are "the same." This can actually be made precise using Definition 2.1. For each $g \in G$, define a function T_g on $\mathbf{X}$ to $\mathbf{X}$ by

$$T_g(x) \equiv F(g, x).$$

Then T_e is the identity function and (ii) simply means

$$T_{g_1}(T_{g_2}(x)) = T_{g_1 g_2}(x).$$

That each T_g is one-to-one and onto is easily verified as is the equation

$$T_{g^{-1}} = T_g^{-1}.$$

Thus $\{T_g | g \in G\}$ is a group under function composition and function composition in this group corresponds to group composition in G.

The reason for the adjective "left" in Definition 2.1, is that there is also a definition of a right group action in which condition (ii) becomes

(ii′) $F(g_1 g_2, x) = F(g_2, F(g_1, x))$.

In these notes, all group actions are defined so that they are left group actions. Some care must be taken in certain examples to insure that a group action is a left group action.

Here is our first example.

EXAMPLE 2.1. Consider $G = \mathrm{Gl}_n$ and $\mathbf{X} = R^n$. Define F on $\mathrm{Gl}_n \times R^n$ by

$$F(g, x) = gx,$$

where gx means the matrix g times the vector x. That F defines a left group action is immediate. In less formal notation, one would say "Gl_n acts on the left of R^n via the group action

$$x \to gx,"$$

where gx means what it did before. Since Gl_n acts on the left of R^n, so does every subgroup of Gl_n—just restrict the action to the subgroup. In particular, O_n and $G_T^+ \subset \mathrm{Gl}_n$ both act on the left of R^n via the action

$$x \to gx. \qquad \square$$

Henceforth we will switch to the less formal description of left group actions and just say something like "consider G acting on the left of $\mathbf{X}$ with action

$$x \to gx."$$

A group G acting on the left of $\mathbf{X}$ induces a natural equivalence relationship among the elements of $\mathbf{X}$, namely x_1 is equivalent to x_2 iff $x_1 = gx_2$ for some $g \in G$. This equivalence relationship divides $\mathbf{X}$ into disjoint subsets called *orbits*, that is

$$O_x = \{gx | g \in G\}$$

is called the *orbit of* x and consists of exactly those elements in $\mathbf{X}$ which are equivalent to x. Hence two points are equivalent iff they are in the same orbit.

EXAMPLE 2.2. Take $\mathbf{X}$ to be the real vector space S_n of all $n \times n$ real symmetric matrices and take $G = O_n$. The left action of O_n on $\mathbf{X}$ is defined by

$$x \to gxg',$$

where g' is the transpose of $g \in O_n$ and gxg' means the product of the three matrices g, x and g'. To describe the orbit of $x \in S_n$, recall the spectral theorem for elements of S_n which asserts that for each x, there is a $g \in O_n$ such that

$$x = g\lambda g',$$

where the diagonal matrix $\lambda \in S_n$ has diagonal elements $\lambda_1 \geq \lambda_2 \geq \cdots \geq \lambda_n$ which are the ordered eigenvalues of x. Thus, two points x_1 and x_2 are equivalent iff x_1 and x_2 have the same vector of ordered eigenvalues. This follows because the group action

$$x \to gxg'$$

does not change the eigenvalues of x and if x_1 is equivalent to x_2, then x_1 and x_2 have the same eigenvalues. Thus, we can say that the eigenvalues provide an index for the orbits in S_n under the action of O_n—an index in the sense that there is a one-to-one correspondence between orbits and vectors (ordered) of eigenvalues. Later in this chapter, we introduce *maximal invariants* which are just functions which provide one-to-one orbit indices. Thus for this example, the vector of eigenvalues provides a maximal invariant. □

EXAMPLE 2.3. For this example, take $\mathbf{X}$ to be $F_{p,n}$ which is the set of $n \times p$ real matrices x which satisfy $x'x = I_p$, the $p \times p$ identity matrix. Thus $x \in F_{p,n}$ iff the p columns of x are the first p columns of some $n \times n$ orthogonal matrix. Let $G = O_n$ which acts on the left of $F_{p,n}$ via

$$x \to gx,$$

where gx means matrix multiplication. Note that $F_{1,n}$ is the sphere of radius 1 on R^n and $F_{n,n} = O_n$. Given $x \in F_{p,n}$, let $g \in O_n$ have as its first p rows the transposes of the first p columns of x. Then the orthogonality of g implies that

$$gx = x_0 = \begin{pmatrix} I_p \\ 0 \end{pmatrix} : n \times p.$$

Thus there is only one orbit in $F_{p,n}$ since every element x is equivalent to x_0. More generally, when G acts on $\mathbf{X}$ and there is only one orbit, we say G acts *transitively* on $\mathbf{X}$. □

EXAMPLE 2.4. Let $\mathscr{L}_{p,n}$ be the vector space of all $n \times p$ real matrices, $p \leq n$, and consider the product group $O_n \times \mathrm{Gl}_p$ whose elements are written (γ, g). The group action is defined by

$$x \to \gamma x g'$$

for $x \in \mathscr{L}_{p,n}$, $\gamma \in O_n$ and $g \in \mathrm{Gl}_p$, where g' means the transpose of g. The reason for the transpose on g is so that the action is a *left* action. Without the transpose, Definition 2.1(ii) does not hold. This action is not transitive, but it "almost" is transitive in the following sense. Let $\mathbf{X} \subset \mathscr{L}_{p,n}$ be all the rank p elements in $\mathscr{L}_{p,n}$ so the complement of $\mathbf{X}$ in $\mathscr{L}_{p,n}$ has Lebesgue measure 0. Then $O_n \times \mathrm{Gl}_p$ acts on $\mathbf{X}$ with the action defined above. To see that this action on $\mathbf{X}$ is transitive, consider $x \in \mathbf{X}$ and write x as

$$x = uv',$$

where $u \in F_{p,n}$ (of Example 2.3) and $v \in G_T^+ \subset \mathrm{Gl}_p$. [This is the so-called *Q-R* decomposition which is usually proved via the Gram–Schmidt orthogonalization procedure; for an example, see Proposition 5.2 in Eaton (1983), page 160.] Hence, for $(\gamma, g) \in O_n \times \mathrm{Gl}_p$,

$$(\gamma, g)x = \gamma x g' = \gamma u v' g' = \gamma u (gv)'.$$

By picking $g = v^{-1}$ and γ so that

$$\gamma u = \begin{pmatrix} I_p \\ 0 \end{pmatrix} \equiv x_0 \in \mathbf{X},$$

it follows that

$$(\gamma, g)x = x_0.$$

Thus, every x is equivalent to x_0, so the group action is transitive. □

The issue of removing a "small" subset from a space in order to make a group action "nicer" in some sense occurs in many examples. In the above example, "small" means a set of Lebesgue measure 0 and "nice" means transitive. However, in other examples, these words can have different meanings.

2.2. Relatively invariant integrals. Consider a space $\mathbf{X}$ (space in the sense introduced in Chapter 1, so $\mathbf{X}$ is a locally compact Hausdorff space whose topology has a countable base) and let G be a topological group.

DEFINITION 2.2. The group G acts *topologically on the left of* $\mathbf{X}$ if G acts on the left of $\mathbf{X}$ and if the action of G, say F: $G \times \mathbf{X} \to \mathbf{X}$, is continuous.

Because all the actions we consider are left actions, the phrase "on the left" is deleted in what follows and we simply say G acts topologically on $\mathbf{X}$. As usual,

$K(\mathbf{X})$ denotes the vector space of continuous functions with compact support defined on $\mathbf{X}$, so integrals are defined on $K(\mathbf{X})$. When G acts topologically on $\mathbf{X}$, then L_g defined on $K(\mathbf{X})$ by

$$(L_g f)(x) = f(g^{-1}x)$$

maps $K(\mathbf{X})$ onto $K(\mathbf{X})$. Note that

$$L_g L_h = L_{gh}$$

so that G acts on the left of $K(\mathbf{X})$ with the action

$$f \to L_g f.$$

DEFINITION 2.3. Let χ be a multiplier for G. An integral J on $K(\mathbf{X})$ is *relatively* (left) *invariant with multiplier* χ if

$$J(L_g f) = \chi(g) J(f)$$

for $f \in K(\mathbf{X})$ and $g \in G$. Equivalently, if J is represented by the Radon measure m, then m is *relatively* (left) *invariant with multiplier* χ if

$$\int f(g^{-1}x) m(dx) = \chi(g) \int f(x) m(dx)$$

for $f \in K(\mathbf{X})$ and $g \in G$.

Here are a couple of examples.

EXAMPLE 2.5. With $\mathbf{X} = R^n$ and $G = \mathrm{Gl}_n$ as in Example 2.1, G acts topologically on $\mathbf{X}$ and Lebesgue measure dx is relatively invariant with multiplier

$$\chi(g) = |\det(g)|. \qquad \square$$

EXAMPLE 2.6. With $\mathbf{X} = R^n$ and $G = O_n$, consider a probability measure P on $\mathbf{X}$ and let $X \in \mathbf{X}$ be a random vector with distribution P—this we write as $\mathscr{L}(X) = P$. Recall that X has a *spherical* distribution if $\mathscr{L}(X) = \mathscr{L}(gX)$ for $g \in O_n$. In terms of P, X is spherical iff P is invariant (with multiplier $\chi \equiv 1$) under O_n, that is, iff

$$P(B) = P(g^{-1}B)$$

for all Borel sets B and $g \in O_n$. If we define the probability measure gP by

$$(gP)(B) = P(g^{-1}B)$$

for all Borel sets B, then X is spherical iff $P = gP$ for $g \in O_n$. Since $\mathscr{L}(X) = P$ implies $\mathscr{L}(gX) = gP$, we have that X is spherical iff P is O_n invariant (i.e., $P = gP$). Notice that the group O_n acts on all the probability measures on R^n via the definition of gP. $\square$

EXAMPLE 2.7. For this example, let S_n be the vector space of $n \times n$ real symmetric matrices and let $G = \mathrm{Gl}_n$ act on S_n via

$$x \to gxg'.$$

With dx denoting Lebesgue measure on S_n, define an integral J by

$$J(f) = \int_{S_n} f(x)\, dx.$$

To see if J is relatively invariant (for some multiplier), consider

$$J(L_g f) = \int_{S_n} f(g^{-1}x(g^{-1})')\, dx$$

and introduce the change of variables $y = g^{-1}x(g^{-1})'$ so $x = gyg'$. This change of variables defines a nonsingular linear transformation on S_n whose determinant is

$$(\det(g))^{n+1}$$

[see Eaton (1983), page 169 for a proof]. Thus

$$dx = |\det(g)|^{n+1}\, dy,$$

so $J(L_g f) = |\det(g)|^{n+1} J(f)$ and J is relatively invariant with the given multiplier. This example is considered again later. □

Example 2.8. As in Example 2.4, let $\mathbf{X} \subset \mathscr{L}_{p,n}$ be all the $n \times p$ real matrices of rank p (so $p \le n$) and take $G = O_n \times \mathrm{Gl}_p$ with the group action defined in Example 2.4. Let dx denote Lebesgue measure restricted to $\mathbf{X}$ and define the integral J by

$$J(f) = \int f(x) \frac{dx}{(\det x'x)^{n/2}} = \int f(x) m(dx)$$

for $f \in K(\mathbf{X})$. We now show J is invariant (relatively invariant with multiplier $\chi \equiv 1$). For $(\gamma, g) \in O_n \times \mathrm{Gl}_p$,

$$J(L_{(\gamma,g)} f) = \int f(\gamma' x (g^{-1})') m(dx)$$

and the change of variable $y = \gamma' x (g^{-1})'$ yields $x = \gamma y g'$. This linear transformation $(y \to \gamma y g')$ on $\mathscr{L}_{p,n}$ has a Jacobian given by

$$|\det(g)|^n$$

[see Eaton (1983), page 168]. Substitution now yields

$$J(L_{(\gamma,g)} f) = \int f(y) |\det(g)|^n \frac{dy}{|\det(g'y'yg')|^{n/2}}$$

$$= \frac{|\det(g)|^n}{|\det gg'|^{n/2}} J(f) = J(f).$$

Thus J is invariant. □

Now we turn to the question of existence and uniqueness of relatively invariant integrals with a given multiplier. Assume G acts topologically on **X** and assume G acts transitively on **X** (this is the only case where we can hope for uniqueness). Fix x_0 in **X** and assume that the function

$$\pi: g \to gx_0$$

on G to **X** is an open mapping (forward images of open sets are open). Note that π maps G onto X because G is assumed to be transitive. Further, let

$$H = \{g | gx_0 = x_0\}$$

so H is a closed subgroup of G. This subgroup is often called the *isotropy subgroup of* x_0. Let Δ_H denote to modulus of H and let Δ denote the modulus of G.

THEOREM 2.1 (Weil). *In order that there exists a relatively invariant integral J with multiplier χ on G, it is necessary and sufficient that χ satisfy the equation*

$$\Delta_H(h) = \chi(h)\Delta(h) \quad \textit{for } h \in H. \tag{2.1}$$

When J exists, it is unique up to a positive constant.

PROOF. See Nachbin (1965), Chapter 3, especially pages 125–141. □

A few remarks are in order concerning Theorem 2.1. The validity of the theorem does not depend on the choice of x_0, that is, if the result is true for one x_0, then it is true for all x_0's. The verification of (2.1) requires the calculation of both the modulus for G and the modulus for H. In the special case when **X** is compact and G is compact, Theorem 2.1 guarantees the existence of a unique G-invariant probability measure on **X**. For future reference, we state this as:

THEOREM 2.2. *Assume the conditions of Theorem* 2.1 *and assume both* **X** *and G are compact. Then there exists a unique G-invariant probability measure on* **X**.

PROOF. Because **X** is compact, all integrals are finite as the function $f \equiv 1$ is integrable. Since G is compact, H is compact so $\Delta = \Delta_H \equiv 1$ and χ must be identically 1 also. Thus, (2.1) holds so there exists a finite G-invariant integral (measure) on **X**. Normalizing this to be a probability measure gives the uniqueness. □

Here are a couple of examples.

EXAMPLE 2.9. With Gl_n acting on R^n, observe that the basic assumption of transitivity of the group action is not satisfied. However Gl_n does act transitively on $R^n - \{0\} \equiv \mathbf{X}$. With

$$x_0 = \begin{pmatrix} 1 \\ 0 \\ \vdots \\ 0 \end{pmatrix} \in R^n,$$

the isotropy subgroup of x_0 is

$$H = \{h | h \in \mathrm{Gl}_n,\ hx_0 = x_0\}.$$

It is easily verified that $h \in H$ iff

$$h = \begin{pmatrix} 1 & b \\ 0 & a \end{pmatrix} \in \mathrm{Gl}_n,$$

where $a \in \mathrm{Gl}_{n-1}$ and b is $1 \times (n-1)$. The modulus of Gl_n is $\Delta \equiv 1$. To compute the modulus of H, a left invariant measure on H is first computed. Let $da\,db$ be Lebesgue measure restricted to H. For $k \in H$, consider

$$J(f) = \int f\left(k^{-1}\begin{pmatrix} 1 & b \\ 0 & a \end{pmatrix}\right) da\,db.$$

Set

$$\begin{pmatrix} 1 & c \\ 0 & d \end{pmatrix} = k^{-1}\begin{pmatrix} 1 & b \\ 0 & a \end{pmatrix},$$

where

$$k = \begin{pmatrix} 1 & \beta \\ 0 & \alpha \end{pmatrix}.$$

Then,

$$\begin{pmatrix} 1 & b \\ 0 & a \end{pmatrix} = k\begin{pmatrix} 1 & c \\ 0 & d \end{pmatrix} = \begin{pmatrix} 1 & \beta \\ 0 & \alpha \end{pmatrix}\begin{pmatrix} 1 & c \\ 0 & d \end{pmatrix} = \begin{pmatrix} 1 & c + \beta d \\ 0 & \alpha d \end{pmatrix}.$$

The Jacobian of this transformation is

$$\phi(k) = |\det(\alpha)|^{n-1}.$$

Thus J is relatively invariant with multiplier ϕ, so from Theorem 1.6,

$$\nu_l(da, db) = \frac{da\,db}{|\det(b)|^{n-1}}$$

is a left-invariant measure on H. Thus, to compute Δ_H, consider

$$J_1(fR_k) = \int f(xk^{-1})\nu_l(dx)$$

with ν_l given above. Here, dx has been written for $da\,db$. The usual change of variable–Jacobian argument yields

$$J_1(fR_k) = |\det(\alpha)| J_1(f).$$

Therefore

$$\Delta_H(k) = |\det(\alpha)|,$$

where

$$k = \begin{pmatrix} 1 & \beta \\ 0 & \alpha \end{pmatrix}.$$

Now consider a multiplier

$$\chi_\delta(g) = |\det(g)|^\delta$$

on G where δ is a fixed real number. Then (2.1) holds iff

$$\Delta_H(h) = \chi_\delta(h) \quad \text{for } h \in H.$$

With

$$h = \begin{pmatrix} 1 & \beta \\ 0 & \alpha \end{pmatrix},$$

$$\chi_\delta(h) = |\det(\alpha)|^\delta,$$

so that (2.1) becomes

$$|\det(\alpha)| = |\det(\alpha)|^\delta$$

for all $\alpha \in \mathrm{Gl}_{n-1}$. Thus δ must be 1 and

$$\chi(g) = |\det(g)|$$

is the only multiplier on Gl_n which satisfies (2.1). Of course, Lebesgue measure on $R^n - \{0\}$ is relatively invariant with multiplier χ and is unique up to a positive constant. □

EXAMPLE 2.10. As in Example 2.3, consider the group O_n acting on $F_{p,n}$. This group action is transitive and both O_n and $F_{p,n}$ are compact sets. Thus there exists a unique O_n-invariant probability distribution on $F_{p,n}$ which we call the *uniform distribution* on $F_{p,n}$. When $p = 1$, this is just the uniform distribution on the sphere of radius 1 in R^n. When $p = n$, then we just get Haar measure on $O_n = F_{n,n}$.

This example is related to the problem of how to define what one means by "a randomly chosen subspace of dimension p in R^n." To see the connection, let $S_{p,n}$ denote all the rank p orthogonal projection matrices defined on R^n. Then $S_{p,n}$ is the image of $F_{p,n}$ under the mapping

$$x \to xx', \qquad x \in F_{p,n}.$$

This map is onto but, of course, not one-to-one. The group O_n acts on $S_{p,n}$ by

$$u \to gug', \qquad u \in S_{p,n},$$

as is suggested by mapping from $F_{p,n}$ to $S_{p,n}$ and the action of O_n on $F_{p,n}$. This action is transitive because for each $u \in S_{p,n}$, there is a $g \in O_n$ such that

$$gug' = \begin{pmatrix} I_p & 0 \\ 0 & 0 \end{pmatrix}.$$

That $S_{p,n}$ is a compact subset of $\mathscr{L}_{n,n}$ is easily established. Thus, there exists a unique O_n-invariant probability on $S_{p,n}$ which we call the *uniform distribution* on $S_{p,n}$ Since there is a one-to-one map between subspaces of dimension p and elements of $S_{p,n}$, this defines a uniform distribution on p dimensional subspaces. □

2.3. Maximal invariants. In this section we consider the problem of calculating explicit representations of orbit indices when a group G acts on a set $\mathbf{X}$. Recall that for $x \in \mathbf{X}$,

$$O_x = \{gx \mid g \in G\}$$

is the *orbit of* x and these orbits are the equivalence class of points which are equivalent under the action of G. That is, x_1 is equivalent to x_2 if $x_1 = gx_2$ for some $g \in G$. The problem is to describe what the orbits are for some examples. This was done in Example 2.2, but here general methods are described along with the examples. The main interest in orbit indices arises from problems in the construction of best invariant tests which is discussed in later chapters.

Let the group G act on $\mathbf{X}$ and suppose a function f maps $\mathbf{X}$ into $\mathbf{Y}$.

DEFINITION 2.4. The function f is *invariant* if $f(x) = f(gx)$. The function f is *maximal invariant* if f is invariant and if $f(x_1) = f(x_2)$ implies $x_1 = gx_2$ for some $g \in G$.

Thus, f is invariant iff f is constant on each orbit in $\mathbf{X}$. Also f is maximal invariant iff f is constant on each orbit and takes different values on different orbits. That is, maximal invariant functions provide an orbit index, namely, knowing the value of maximal invariant f at x is equivalent to knowing O_x. Notice that the image space $\mathbf{Y}$ plays no role in Definition 2.4. In examples, $\mathbf{Y}$ is ordinarily chosen with convenience in mind.

THEOREM 2.3. *Suppose f: $\mathbf{X} \to \mathbf{Y}_1$ is maximal invariant under the action of G on $\mathbf{X}$. Then a function h: $\mathbf{X} \to \mathbf{Y}_2$ is invariant iff there exists a function k, k: $\mathbf{Y}_1 \to \mathbf{Y}_2$, such that $h(x) = k(f(x))$.*

PROOF. If k exists, obviously h is invariant since f is invariant. Conversely, suppose h is invariant and define k on $\mathbf{Y}_1$ as follows:

(i) If $y \in \mathbf{Y}_1$ is given by $y = f(x)$, set $k(y) = h(x)$.
(ii) If $y \in \mathbf{Y}_1$ is not in the range of f, define k arbitrarily.

That k is well defined follows from the invariance of h and the maximal invariance of f. Obviously $h(x) = k(f(x))$. □

This result shows that once a maximal invariant is known, then all the invariant functions are known, namely, they are just the functions of a maximal invariant.

The first method we use to construct a maximal invariant might be termed "the reduction method." The idea is to "reduce" or "transform" a point x to a canonical form via elements of G is such a way as to pick out a particular point from each orbit.

EXAMPLE 2.11. Consider the group O_n acting on R^n, in the usual way. Given $x \in R^n$, there exists a $g \in O_n$ such that

$$gx = \|x\| x_0,$$

where $\|x\|$ is the length of x and x_0 is the first standard unit vector

$$x_0 = \begin{pmatrix} 1 \\ 0 \\ \vdots \\ 0 \end{pmatrix}.$$

For example, take g to have first row equal to $x'/\|x\|$. That the function

$$f(x) = \|x\| x_0$$

is maximal invariant is proved as follows. Clearly f is invariant since x and gx have the same length. Now $f(x) = f(y)$ iff $\|x\| = \|y\|$. Thus when $\|x\| = \|y\|$, we must find a $g \in G$ so that $gx = y$. Pick g_1 and g_2 such that

$$g_1 x = \|x\| x_0 = \|y\| x_0 = g_2 y.$$

Then $y = g_2^{-1} g_1 x$, so x and y are in the same orbit. Clearly, any one-to-one function of $\|x\|$ is also a maximal invariant, and a function is invariant iff it can be written as a function of $\|x\|$. □

EXAMPLE 2.12. As in Example 2.2, let $G = O_n$ act on S_n with action

$$x \to gxg', \qquad g \in O_n.$$

Given $x \in S_n$, the spectral theorem shows there is a $g \in O_n$ such that

$$gxg' = \lambda(x),$$

where $\lambda(x)$ is an $n \times n$ diagonal matrix with diagonal elements $\lambda_1(x) \geq \lambda_2(x) \geq \cdots \geq \lambda_n(x)$, which are the ordered eigenvalues of x. The claim is that

$$f(x) = \lambda(x)$$

is maximal invariant. Since eigenvalues are invariant under the group action $x \to gxg'$, f is invariant. If $f(x_1) = f(x_2)$, we must show that $x_2 = gx_1g'$ for some $g \in G$. Pick $g_1, g_2 \in G$ such that

$$g_i x_i g_i' = \lambda(x_i), \qquad i = 1, 2.$$

With $g = g_2' g_1$,

$$gx_1g' = g_2' g_1 x_1 g_1' g_2 = g_2' \lambda(x_1) g_2 = g_2' \lambda(x_2) g_2 = x_2.$$

Thus the vector of ordered eigenvalues of x is a maximal invariant. □

EXAMPLE 2.13. For this example the product group $O_n \times O_p$ acts on $\mathscr{L}_{p,n}$ via

$$x \to gxh', \qquad g \in O_n, \ h \in O_p,$$

where we assume $p \leq n$. The reduction argument here follows from the singular value decomposition theorem [for the version used here, see Anderson (1984), page 590]. According to this theorem, given $x \in \mathscr{L}_{p,n}$, there exists $g \in O_n$ and $h \in O_p$ such that

$$x = g\lambda(x)h',$$

where $\lambda(x)$: $n \times p$ has the form

$$\lambda(x) = \begin{pmatrix} \lambda_1 & & & 0 \\ & \lambda_2 & & \\ 0 & & \ddots & \\ & & \cdots & \lambda_p \\ & & 0 & \end{pmatrix}$$

and $\lambda_1 \geq \cdots \geq \lambda_p \geq 0$ are the square roots of the eigenvalues of $x'x$. (The numbers $\lambda_1, \ldots, \lambda_p$ are often called the *singular values* of x.) The claim is that

$$f(x) = \lambda(x)$$

is maximal invariant. Obviously f is invariant since singular values do not change under the given group action. If $f(x_1) = f(x_2)$ [so $\lambda(x_1) = \lambda(x_2)$], pick $g_1, g_2 \in O_n$ and $h_1, h_2 \in O_p$ such that

$$g_i x_i h_i' = \lambda(x_i), \qquad i = 1, 2.$$

Since $\lambda(x_1) = \lambda(x_2)$, this implies that

$$x_2 = g_2' g_1 x_1 h_1' h_2,$$

so x_1 and x_2 are in the same orbit. Thus f is maximal invariant. □

The next method of constructing maximal invariants involves finding a function τ mapping $\mathbf{X}$ into G, which has the property

$$\tau(gx) = g\tau(x), \tag{2.2}$$

where $g\tau(x)$ means the composition of the two group elements g and $\tau(x)$. Temporarily assume we can find such a τ and consider

$$f(x) = (\tau(x))^{-1}x,$$

where $(\tau(x))^{-1}$ is the inverse in G of $\tau(x)$. Using (2.2),

$$\begin{aligned} f(gx) &= (\tau(gx))^{-1}gx = (g\tau(x))^{-1}gx \\ &= (\tau(x)^{-1})g^{-1}gx = (\tau(x))^{-1}x = f(x), \end{aligned}$$

so f is invariant. To show f is a maximal invariant, suppose $f(x_1) = f(x_2)$. Then

$$(\tau(x_1))^{-1}x_1 = (\tau(x_2))^{-1}x_2,$$

which yields

$$x_2 = \left[\tau(x_2)(\tau(x_1))^{-1}\right]x_1 = gx_1.$$

This shows x_1 and x_2 are in the same orbit which entails the maximal invariance of f. Here are some examples where τ can be constructed.

EXAMPLE 2.14. With $G = R^1$ and $\mathbf{X} = R^n$, consider the action

$$x \to x + ge_n, \qquad x \in R^n,$$

where e_n is the vector of 1's in R^n and $g \in R^1$. Define τ by

$$\tau(x) = \bar{x},$$

where as usual $\bar{x} = n^{-1}\Sigma_1^n x_i$. An easy calculation shows (2.2) holds and thus

$$f(x) = (\tau(x))^{-1}x = x - \bar{x}e_n$$

is a maximal invariant. □

EXAMPLE 2.15. For this example elements of G consist of pairs (a, b) with $a > 0$ and $b \in R^1$ and the group operation is

$$(a_1, b_1)(a_2, b_2) = (a_1a_2, a_1b_2 + b_1).$$

With e_n denoting the vector of 1's in R^n, the set $\mathbf{X}$ is

$$\mathbf{X} = R^n - \operatorname{span}\{e_n\},$$

so $\mathbf{X}$ is R^n with a line removed. The action of G on $\mathbf{X}$ is

$$(a, b)x = ax + be_n.$$

For $x \in \mathbf{X}$, $\bar{x} \in R^1$ is as in the last example and

$$s(x) = \left[\sum_1^n (x_i - \bar{x})^2\right]^{1/2}.$$

Define τ on $\mathbf{X}$ to G by

$$\tau(x) = (s(x), \bar{x}).$$

Note that $s(x) > 0$ because the line where s vanishes has been removed from R^n. (This is another example of removing a "small" set.) An easy calculation shows that

$$\tau((a, b)x) = (a, b)(s(x), \bar{x}).$$

Since (2.2) holds,

$$f(x) = (\tau(x))^{-1}x = \frac{x - \bar{x}e_n}{s(x)}$$

is maximal invariant. □

In some examples, it is possible to calculate a maximal invariant by decomposing the group into subgroups and doing the calculation for each subgroup, that is, doing the calculation in steps. The applicability of this stepwise procedure depends on the notion of moving a group action from one space to another. Here is the idea. Suppose G acts on $\mathbf{X}$ and a fixed function f maps $\mathbf{X}$ onto $\mathbf{Y}$. In order to try to define a group action on $\mathbf{Y}$, one possibility is to write $y = f(x)$ and then define gy by

$$gy = f(gx). \tag{2.3}$$

Conditions under which this gives an unambiguous definition are provided by:

THEOREM 2.4. *Suppose that $f(x_1) = f(x_2)$ implies that $f(gx_1) = f(gx_2)$ for all $g \in G$. Then* (2.3) *defines G acting on $\mathbf{Y}$.*

PROOF. To see that gy is well defined, if $y = f(x_1) = f(x_2)$, then by assumption $f(gx_1) = f(gx_2)$ for all $g \in G$. Thus $gy = f(gx_1)$ for all $g \in G$ is an unambiguous specification of gy. Obviously, $ey = y$ for all $y \in \mathbf{Y}$. Also for $y = f(x)$ and $g_1, g_2 \in G$,

$$\begin{aligned} g_1(g_2 y) &= g_1 f(g_2 x) = f(g_1 g_2 x) \\ &= (g_1 g_2) f(x) = (g_1 g_2) y. \end{aligned}$$

Thus G acts on $\mathbf{Y}$. □

To describe the stepwise calculation of a maximal invariant, suppose G acts on $\mathbf{X}$ and suppose H and K are subgroups of G which *generate* G, that is, each $g \in G$ can be written in the form $g = h_1 k_1 h_2 k_2 \cdots h_r k_r$ for some integer r where $h_i \in H$ and $k_i \in K$, $i = 1, \ldots, r$.

THEOREM 2.5. *Suppose that under the action of H on $\mathbf{X}$, the function f_1 mapping $\mathbf{X}$ onto $\mathbf{Y}$ is a maximal invariant and satisfies*

$$f_1(x_1) = f_1(x_2) \quad \textit{implies} \quad f_1(kx_1) = f_1(kx_2)$$

for all $k \in K$. Also suppose that f_2 mapping $\mathbf{Y}$ into $\mathbf{Z}$ is a maximal invariant under the induced action of K acting on $\mathbf{Y}$ (as described in Theorem 2.4). *Then $f(x) = f_2(f_1(x))$ is a maximal invariant under the action of G on $\mathbf{X}$.*

PROOF. Recall that the action of K on $\mathbf{Y}$ is defined by: write $y = f_1(x)$ and set $ky = f_1(kx)$. To show f is invariant, consider $h \in H$. Then

$$f(hx) = f_2(f_1(hx)) = f_2(f_1(x))$$

since f_1 is H invariant. For $k \in K$,

$$f(kx) = f_2(f_1(kx)) = f_2(kf_1(x)) = f_2(f_1(x))$$

by definition of $kf_1(x)$ and the invariance of f_2. It is now an easy induction argument to show that

$$f(h_1 k_1 h_2 k_2 \cdots h_r k_r x) = f(x)$$

for all $r = 1, 2, \ldots$ and $h_i \in H$, $k_i \in K$. Thus $f(gx) = f(x)$ since H and K generate G. To show f is maximal, suppose $f(x_1) = f(x_2)$. Hence

$$f(x_1) = f_2(f_1(x_1)) = f_2(f_1(x_2))$$

which, by the maximality of f_2 implies there is a $k \in K$ so that

$$kf_1(x_1) = f_1(x_2) = f_1(kx_1).$$

The final equality follows from the definition of the action of K on $\mathbf{Y}$. The maximality of f_1 implies there is an $h \in H$ so that

$$x_2 = h(kx_1) = (hk)x_1.$$

Thus, $x_2 = gx_1$ for some g so x_1 and x_2 are in the same orbit. Thus f is maximal. □

The advantages of reducing in steps are most apparent when dealing with rather complicated problems. The next example, in which the correlation coefficient is a maximal invariant, well illustrates this situation.

EXAMPLE 2.16. With $n \geq 3$, let e_n denote the vector of 1's in R^n and set

$$Q = I_n - n^{-1}e_n e_n'.$$

Clearly, Q is the orthogonal projection onto the subspace perpendicular to e_n. Let $\mathbf{X} \subset \mathcal{L}_{2,n}$ be the set of $n \times 2$ real matrices x such that Qx has rank 2. Note that

$$s(x) = x'Qx$$

is the (unnormalized) sample covariance matrix when one observes n two dimensional vectors and arranges them into x: $n \times 2$ whose rows are the transposes of the data vectors. Points in $\mathbf{X}$ are those sample points such that $s(x)$ has rank 2, that is, $s(x)$ is positive definite.

The group for this example is a product group $G = G_1 \times G_2$ with

$$G_1 = \{\gamma | \gamma \in O_n, \gamma e_n = e_n\}.$$

The group G_2 is a subgroup of Al_2. An element of G_2 is a pair (a, b) with $b \in R^2$ and

$$a = \begin{pmatrix} a_1 & 0 \\ 0 & a_2 \end{pmatrix} \in \mathrm{Gl}_2, \qquad a_i > 0,\ i = 1,2.$$

The action of an element $(\gamma, (a, b))$ in $G_1 \times G_2$ on $x \in \mathcal{L}_{2,n}$ is

$$(\gamma, (a, b))x = \gamma x a' + e_n b',$$

where the prime denotes transpose. Even though $a' = a$, we write a' to remind the reader that transposes are necessary in such situations to insure that actions are indeed left actions. That we have a left action is easy to check, as is the fact that points in $\mathbf{X}$ are mapped into points in $\mathbf{X}$.

To construct a maximal invariant, consider the two subgroups H and K defined as follows. With $e \in \mathrm{Gl}_2$ denoting the 2×2 identity matrix and $I \in O_n$ denoting the $n \times n$ identity matrix, let

$$H = \{(\gamma, (e, b)) | \gamma \in G_1, (e, b) \in G_2\}$$

and

$$K = \{(I, (a, 0)) | (a, 0) \in G_2\}.$$

Since every element $(\gamma,(a, b))$ in G can be expressed as

$$(\gamma,(a,b)) = (\gamma,(e,b))(I,(a,0)),$$

H and K generate G.

Our first claim is that f_1 defined on $\mathbf{X}$ onto S_2^+ (the set of 2×2 positive definite matrices) by

$$f_1(x) = s(x)$$

is maximal invariant under the action of H on $\mathbf{X}$. To see that f_1 is H-invariant, consider $(\gamma,(e, b)) \in H$ so

$$(\gamma,(e,b))x = \gamma x + e_n b'.$$

Thus

$$Q(\gamma,(e,b))x = Q\gamma x = \gamma Q x$$

because $Qe_n = 0$ and Q commutes with each $\gamma \in G_1$. Because $Q = Q' = Q^2$,

$$\begin{aligned} f_1((\gamma,(e,b))x) &= [(\gamma,(e,b))x]'Q[(\gamma,(e,b))x] \\ &= (\gamma Qx)'(\gamma Qx) = x'Qx = f_1(x) \end{aligned}$$

and hence f_1 is invariant. For the maximality of f_1, suppose

$$f_1(x) = x'Qx = f_1(y) = y'Qy$$

with x, $y \in \mathbf{X}$. Since $Q^2 = Q = Q'$,

$$(Qx)'Qx = (Qy)'Qy,$$

which implies there is a $\gamma \in G_1$ such that

$$\gamma Qy = Qx.$$

The existence of this γ follows from a minor modification of Proposition 1.20 in Eaton (1983). From the definition of Q we have

$$x = x - e_n\bar{x}' + e_n\bar{x}' = Qx + e_n\bar{x}',$$

where $\bar{x} = x'e_n/n \in R^2$. Therefore,

$$\begin{aligned} x &= Qx + e_n\bar{x}' = \gamma Qy + e_n\bar{x}' = \gamma(y - e_n\bar{y}') + e_n\bar{x}' \\ &= (\gamma,(0,\bar{x} - \bar{y}))y. \end{aligned}$$

Hence x and y are in the same H orbit, so f_1 is maximal.

To apply Theorem 2.5, it must be verified that

$$f_1(x) = f_1(y) \quad \text{implies} \quad f_1(kx) = f_1(ky)$$

for all $k \in K$. With

$$k = \begin{pmatrix} a_1 & 0 \\ 0 & a_2 \end{pmatrix} \in K,$$

it is an easy calculation to verify that

$$s(kx) = ks(x)k'.$$

Thus $f_1(x) = f_1(y)$ implies that

$$f_1(kx) = s(kx) = ks(x)k' = ks(y)k' = s(ky) = f(ky),$$

so Theorem 2.5 applies.

Finally, a maximal invariant for the action of K on S_2^+ given by

$$s \to ksk', \qquad k \in K,$$

needs to be found. Writing

$$s = \begin{pmatrix} s_{11} & s_{12} \\ s_{21} & s_{22} \end{pmatrix} \in S,$$

an easy argument shows that

$$f_2(s) = \frac{s_{12}}{\sqrt{s_{11}s_{22}}}$$

is a maximal invariant on S_2^+. Hence by Theorem 2.5,

$$f(x) = \frac{s_{12}(x)}{\sqrt{s_{11}(x)s_{22}(x)}}$$

is a maximal invariant under the action of G on **X**. Of course, $f(x)$ is just the sample correlation coefficient. □

2.4. Induced group actions: Equivariance. The situation described in Theorem 2.4 is one circumstance in which it is possible to induce a group action on one space, given a group action on anther space. In this section other circumstances are discussed where group actions can be induced on a space, based on a group action on an associated space. Applications of these ideas occur throughout these notes.

First, suppose that a group G acts on **X** and let f be a function on **X** to **Y**. Here is a useful definition of a new function, denoted by gf for $g \in G$, which also maps **X** to **Y**:

$$(gf)(x) = f(g^{-1}x). \tag{2.4}$$

It is clear that $ef = f$ where e is the identity in G and

$$g_1(g_2 f) = (g_1 g_2)f, \tag{2.5}$$

that is

$$\begin{aligned}(g_1(g_2 f))(x) &= (g_2 f)(g_1^{-1}x) = f(g_2^{-1}g_1^{-1}x) = f((g_1g_2)^{-1}x)\\ &= ((g_1g_2)f)(x).\end{aligned}$$

Of course, the reason for the inverse in the definition (2.4) is so that (2.5) is valid. Hence if **Z** is a set of functions from **X** to **Y** such that $f \in$ **Z** implies $gf \in$ **Z** for all $g \in G$, then G acts on **Z** via (2.4). This group action on the set of functions **Z** should not be confused with the group action induced on **Y** which is described in Theorem 2.4.

EXAMPLE 2.17. Take **X** $= \{1, 2, \ldots, n\}$ and let G be the group of all one-to-one functions from **X** to **X** (the permutation group of **X**). Let **Z** be the set of all real valued functions defined on **X**. A point in **Z** can be thought of as a vector in

R^n, namely, $f \in \mathbf{Z}$ corresponds to a vector in R^n given by

$$\begin{pmatrix} f(1) \\ f(2) \\ \vdots \\ f(n) \end{pmatrix} \in R^n.$$

Conversely, every point in R^n is a point in $\mathbf{Z}$. According to (2.4), if $f \in R^n$ has ith coordinate $f(i)$, then gf has ith coordinate $(gf)(i) = f(g^{-1}i)$. This definition yields the left group action of the permutation group G on R^n. There is some confusion in the literature concerning the action of the permutation group on R^n. □

Example 2.18. Consider a measurable space $(\mathbf{X}, \mathscr{B})$ and suppose that G acts on $\mathbf{X}$ in such a way that each g is a bimeasurable mapping. In this circumstance, we say that G acts *measurably* on $(\mathbf{X}, \mathscr{B})$. Thus, G acts on the σ-algebra $\mathscr{B}$ in the obvious way:

$$gB = \{x | x = gy,\ y \in B\}.$$

Now, let P be a probability measure and think of P as a map from $\mathscr{B}$ into $[0, 1]$. Then, according to (2.4) (with $\mathbf{X} = \mathscr{B}$ and $f = P$),

$$(gP)(B) = P(g^{-1}B).$$

In other words, gP is a probability measure on $\mathscr{B}$ whose value at B is $P(g^{-1}B)$.

In terms of random variables, the above means that if $\mathscr{L}(X) = P$, then $\mathscr{L}(gX) = gP$ for X taking values in $\mathbf{X}$. To see this, first observe that $\mathscr{L}(X) = P$ means

$$\text{Prob}(X \in B) = P(B).$$

Therefore,

$$\text{Prob}(gX \in B) = \text{Prob}(X \in g^{-1}B) = P(g^{-1}B) = (gP)(B),$$

so $\mathscr{L}(gX) = gP$. This definition of gP appeared in Example 2.10. □

Example 2.19. This example deals with induced group actions for randomized decision functions (also known as Markov kernels, transition probability kernels, among other things). Consider two measurable spaces $(\mathbf{X}, \mathscr{B}_1)$ and $(\mathbf{Y}, \mathscr{B}_2)$. A *Markov kernel* δ is a function defined on $\mathscr{B}_2 \times \mathbf{X}$ into $[0, 1]$ such that:

(i) For each $x \in \mathbf{X}$, $\delta(\cdot | x)$ is a probability measure on $\mathscr{B}_2$.
(ii) For each $B \in \mathscr{B}_2$, $\delta(B | \cdot)$ is a $\mathscr{B}_1$ measurable function.

Suppose that the group G acts measurably on both $(\mathbf{X}, \mathscr{B}_1)$ and $(\mathbf{Y}, \mathscr{B}_2)$ so G acts on $\mathscr{B}_2 \times \mathbf{X}$ via

$$g(B, x) = (gB, gx).$$

If δ is Markov kernel, then according to (2.4) (with $\mathbf{X} = \mathbf{B}_2 \times \mathbf{X}$ and $f = \delta$) G

acts on δ via

$$(g\delta)(B|x) = \delta(g^{-1}B|g^{-1}x).$$

This example occurs later in the discussion of invariance and decision theory. □

Yet another method for inducing a group action concerns what might be called the "kernel method." Suppose G acts on $\mathbf{X}$ and $K(x, y)$ is a function of two variables defined on $\mathbf{X} \times \mathbf{Y}$. The range space of the kernel K is not relevant. The idea here is to try to use K to induce a group action on $\mathbf{Y}$ in such a way that K becomes an invariant function. To say it another way, when can a group action on $\mathbf{Y}$ be specified so that $K(x, y) = K(gx, gy)$ for all x, y, g?

THEOREM 2.6. *Consider G acting on $\mathbf{X}$ and suppose K is defined on $\mathbf{X} \times \mathbf{Y}$. Assume that for each $g \in G$ and each $y \in \mathbf{Y}$, there exists a unique $y' \in \mathbf{Y}$ such that*

$$K(gx, y') = K(x, y) \tag{2.6}$$

for all $x \in \mathbf{X}$. Then G acts on $\mathbf{Y}$ via the defined group action $gy = y'$. With this group action,

$$K(gx, gy) = K(x, y),$$

so K is an invariant function on $\mathbf{X} \times \mathbf{Y}$.

PROOF. That $ey = y$ for $y \in Y$ is clear by the uniqueness of y'. To verify we have a left action, consider $(g_1g_2)y$ and use (2.6) to compute as follows:

$$K((g_1g_2)x, (g_1g_2)y) = K(x, y) \quad \text{for } x \in \mathbf{X}$$

implies that

$$K(z, (g_1g_2)y) = K(g_2^{-1}(g_1^{-1}z), y) = K(g_1^{-1}z, g_2y) = K(z, g_1(g_2y))$$

for all $z \in \mathbf{X}$. Again uniqueness implies

$$(g_1g_2)y = g_1(g_2y).$$ □

EXAMPLE 2.20. Consider a parametric family of probability measures $\mathscr{P} = \{P(\cdot|\theta)|\theta \in \Theta\}$, defined on a measurable space $(\mathbf{X}, \mathscr{B})$. Suppose that G acts measurably on $(\mathbf{X}, \mathscr{B})$. The family $\mathscr{P}$ is *G-invariant* if $P \in \mathscr{P}$ implies $gP \in \mathscr{P}$ for all $P \in \mathscr{P}$ and $g \in G$. Here the notation gP of Example 2.18 is used.

Assuming the family $\mathscr{P}$ is G-invariant, also assume that

$$P(B|\theta_1) = P(B|\theta_2) \quad \text{for all } B \in \mathscr{B}$$

implies that $\theta_1 = \theta_2$, that is, the points in Θ are in one-to-one correspondence with the elements of the family $\mathscr{P}$. To apply Theorem 2.6, define K on $\mathbf{B} \times \Theta$ by

$$K(B, \theta) = P(B|\theta).$$

To show that the assumptions on K assumed in Theorem 2.6 hold, consider $\theta \in \Theta$ and $g \in G$. Since $\mathscr{P}$ is invariant, $P(g^{-1}B|\theta) = (gP(\cdot|\theta))(B)$ is in $\mathscr{P}$, so there exists a $\theta' \in \Theta$ such that

$$(gP(\cdot|\theta))(B) = P(B|\theta')$$

for all $B \in \mathscr{B}$. In terms of K, this means that there is a θ' such that

$$K(B, \theta') = K(g^{-1}B, \theta)$$

for all B and by the assumption on the family $\mathscr{P}$, θ' is unique. Theorem 2.6 implies that the natural induced group action on Θ yields

$$P(gB|g\theta) = P(B|\theta)$$

for $g \in G$, $B \in \mathscr{B}$ and $\theta \in \Theta$. Thus, if $\mathscr{L}(X) = P(\cdot|\theta)$, then $\mathscr{L}(gX) = P(\cdot|g\theta)$ because

$$\mathscr{L}(gX) = gP(\cdot|\theta).$$

This example is treated more completely in the next lecture. □

Finally, the notation of an equivariant function is introduced. A special case of this notion arose in the construction of a maximal invariant via the function τ in Equation (2.2). For the general case, suppose a group G acts on both $\mathbf{X}$ and $\mathbf{Y}$.

DEFINITION 2.5. A function f on $\mathbf{X}$ to $\mathbf{Y}$ is *equivariant* if

$$f(gx) = gf(x) \quad \text{for } g \in G,\ x \in \mathbf{X}. \tag{2.7}$$

The terminology in the statistical literature is not consistent. In some works, condition (2.5) is called invariance, but recently the tendency has been to the word equivariance. Note that when the group action of G on $\mathbf{Y}$ is trivial (that is, $gy = y$ for all g and all y), then equivariance reduces to invariance.

Given G acting on $\mathbf{X}$ and $\mathbf{Y}$, it seems rather difficult to give a description of all the equivariant functions. However, given G acting on $\mathbf{X}$ and given a function f, the results of Theorem 2.4 give the necessary and sufficient condition for the existence of a group action on $\mathbf{Y}$ such that (2.7) holds. In fact the condition of Theorem 2.4,

$$f(x_1) = f(x_2) \text{ implies that } f(gx_1) = f(gx_2) \text{ for all } g \in G, \tag{2.8}$$

is precisely the necessary and sufficient condition that f be equivariant according to (2.7). Theorem 2.4 establishes the implication in one direction. That (2.7) implies (2.8) is obvious.

Equivariant functions arise naturally in estimation problems which are invariant under a group (these are discussed in detail later). Here are some examples which are related to estimation problems.

EXAMPLE 2.21. Take $\mathbf{X} = \mathscr{L}_{p,n}$ as the vector space of $n \times p$ real matrices and $\mathbf{Y} = S_p$ as the vector space of $p \times p$ real symmetric matrices. The group Gl_p acts on $\mathbf{X}$ by

$$x \to g(x) = xg', \qquad g \in \mathrm{Gl}_p$$

and Gl_p acts on S_p by

$$y \to g(y) = gyg', \qquad g \in \mathrm{Gl}_p.$$

Fix an $n \times n$ symmetrix matrix B and define f by

$$f(x) = x'Bx.$$

Then

$$f(g(x)) = f(xg') = gx'Bxg' = g(f(x))$$

so f is equivariant. □

EXAMPLE 2.22. For this example, G is the group G_T^+ of $p \times p$ lower triangular matrices with positive diagonal elements, $\mathbf{X}$ is the set of all $n \times p$ real matrices of rank p and $\mathbf{Y}$ is G_T^+. Recall that each $x \in \mathbf{X}$ can be written uniquely as

$$x = \gamma g',$$

where γ is an $n \times p$ matrix which satisfies $\gamma'\gamma = I_p$ and $g \in G_T^+$ [see Proposition 5.2 in Eaton (1983)]. Define f on $\mathbf{X}$ to G_T^+ by $f(x)$ is the unique element in G_T^+ such that

$$x = \gamma(f(x))'$$

as above. With G_T^+ acting on $\mathbf{X}$ by

$$x \to g(x) = xg'$$

and with G_T^+ acting on itself via left multiplication

$$h \to gh, \qquad h \in G_T^+,$$

it is easily verified that

$$f(g(x)) = gf(x).$$

Thus f is equivariant. □

Equivariant functions can, under certain conditions, be used in conjunction with Haar measure arguments to define invariant integrals.

EXAMPLE 2.23. Let S_p^+ be the set of positive definite matrices and let G_T^+ be the group of $p \times p$ lower triangular matrices with positive diagonal elements. The function ϕ on G_T^+ to S_p^+ defined by

$$\phi(h) = hh'$$

is one-to-one, onto, bicontinuous (a homeomorphism) and is equivariant,

$$\phi(gh) = ghh'g' = g(hh'),$$

where G_T^+ acts on S_p^+ in the usual way:

$$x \to g(x) = gxg'.$$

The group G_T^+ acts transitively on S_p^+ and Theorem 2.1 together with Example 2.7 shows that the integral

$$J_1(f) = \int_{S_p^+} f(x) \frac{dx}{(\det(x))^{(p+1)/2}}$$

is invariant under this group action.

However, consider the integral

$$J_2(f) = \int_{G_T^+} f(\phi(h))\nu_l(dh) = \int f(hh')\nu_l(dh)$$

for $f \in K(S_p^+)$, where ν_l is a left-invariant measure on G_T^+. The equivariance of ϕ shows that

$$J_2(L_g f) = \int f(g^{-1}(\phi(h)))\nu_l(dh) = \int f(\phi(g^{-1}h))\nu_l(dh) = J_2(f)$$

for $g \in G_T^+$ and $f \in K(S_p^+)$. The uniqueness assertion of Theorem 2.1 shows there is a fixed constant $c > 0$ such that

$$\int f(x)\frac{dx}{(\det(x))^{(p+1)/2}} = c\int f(hh')\nu_l(dh)$$

for all $f \in K(S_p^+)$ and hence for all f which are integrable. The value of the constant c depends on the explicit choice for ν_l. With the choice

$$\nu_l(dh) = \frac{dh}{\prod_{i=1}^p h_{ii}^i}$$

as in Example 1.10, the constant c is 2^p. This is proved by choosing

$$f(x) = |\det(x)|^r \exp[-\tfrac{1}{2}\operatorname{tr} x]$$

and evaluating the two integrals for some convenient choice of the number r. □

CHAPTER 3

Invariant Statistical Models

In this lecture, invariant statistical models are introduced and a variety of examples is given. Invariant testing problems and equivariant estimators are introduced. Univariate and multivariate linear models provide a host of standard examples.

3.1. Invariant models. Given a measurable space $(\mathbf{X}, \mathscr{B})$, a family of probability measures $\mathscr{P}$ defined on $\mathscr{B}$ is a *statistical model*. If the random variable X takes values in $\mathbf{X}$ and $\mathscr{L}(X) \in \mathscr{P}$, then we say $\mathscr{P}$ is a model for $\mathscr{L}(X)$.

DEFINITION 3.1. Suppose the group G acts measurably on $\mathbf{X}$. The statistical model $\mathscr{P}$ is *G-invariant* if for each $P \in \mathscr{P}$, $gP \in \mathscr{P}$ for all $g \in G$.

When the model $\mathscr{P}$ is G-invariant, G acts on $\mathscr{P}$ according to Definition 2.1. Further, when $\mathscr{P}$ is a model for $\mathscr{L}(X)$, then $\mathscr{P}$ is G-invariant means that $\mathscr{L}(gX) \in \mathscr{P}$ whenever $\mathscr{L}(X) \in \mathscr{P}$ for all $g \in G$.

EXAMPLE 3.1. Consider $\mathbf{X} = R^n$ with the Borel σ-algebra and let $f_0(\|x\|^2)$ be any probability density on R^n. Then the probability measure P_0 defined by

$$P_0(B) = \int I_B(x) f_0(\|x\|^2)\, dx$$

is *orthogonally invariant* as described in Example 2.6. That is, for each $g \in O_n$, the group of $n \times n$ orthogonal matrices, $gP_0 = P_0$. Thus $\mathscr{P} = \{P_0\}$ is O_n-invariant.

Of course, there are other orthogonally invariant probabilities than those defined by such a density. In fact, given any probability measure Q on R^n, define P by

$$P(B) = \int_{O_n} (gQ)(B)\nu(dg) = \int Q(g^{-1}B)\nu(dg),$$

where ν is invariant probability measure on O_n. Clearly P is O_n-invariant. Thus, averaging gQ over g with respect to the Haar measure on O_n always produces an O_n-invariant probability.

It is easy to see that this procedure is valid for any compact group G acting measurably on a space $(\mathbf{X}, \mathscr{B})$. That is, let Q be a probability on $(\mathbf{X}, \mathscr{B})$ and define P by

$$P = \int gQ\nu(dg),$$

where ν is the Haar probability measure on G. The above equation means

$$P(B) = \int (gQ)(B)\nu(dg)$$

for $B \in \mathscr{B}$. That P is G-invariant follows from

$$(hP)(B) = P(h^{-1}B) = \int (gQ)(h^{-1}B)\nu(dg) = \int Q(g^{-1}h^{-1}B)\nu(dg)$$

$$= \int Q\big((hg)^{-1}B\big)\nu(dg) = \int Q(g^{-1}B)\nu(dg) = P(B). \qquad \square$$

More detail concerning the structure of probabilities invariant under compact groups is given in the next lecture. Here is a standard parametric example where the group is not compact.

EXAMPLE 3.2. Consider $X_1, \ldots, X_n$ iid $N(\mu, \sigma^2)$ on R^1. Then the random vector

$$X = \begin{pmatrix} X_1 \\ \vdots \\ X_n \end{pmatrix} \in R^n$$

has the distribution

$$\mathscr{L}(X) = N_n\big(\mu e_n, \sigma^2 I_n\big),$$

where e_n is the vector of 1's in R^n. A statistical model for $\mathscr{L}(X)$ is

$$\mathscr{P} = \big\{N\big(\mu e_n, \sigma^2 I_n\big) | \mu \in R^1, \sigma^2 > 0\big\}.$$

The appropriate group G for this example has group elements which are triples (γ, a, b) with $a > 0$, $b \in R^1$ and $\gamma \in O_n$ such that $\gamma e_n = e_n$. The group action on R^n is

$$x \to a\gamma x + be_n$$

and the group operation is

$$(\gamma_1, a_1, b_1)(\gamma_2, a_2, b_2) = (\gamma_1\gamma_2, a_1 a_2, a_1 b_2 + b_1).$$

That $\mathscr{P}$ is G-invariant follows from the observation that when

$$\mathscr{L}(X) = N\big(\mu e_n, \sigma^2 I_n\big),$$

then

$$\mathscr{L}((\gamma, a, b)X) = \mathscr{L}(a\gamma X + be_n) = N\big((a\mu + b)e_n, a^2\sigma^2 I_n\big)$$

which is an element of the model $\mathscr{P}$. $\square$

Example 3.2 is one instance of a standard method of generating an invariant model. Given a fixed probability Q_0 on $(\mathbf{X}, \mathscr{B})$ and a group G acting measurably on $(\mathbf{X}, \mathbf{B})$, let

$$\mathscr{P} = \{gQ_0 | g \in G\}.$$

Obviously $\mathscr{P}$ is G-invariant because $h(gQ_0) = (hg)Q_0$ which is in $\mathscr{P}$ for $gQ_0 \in \mathscr{P}$. In the previous example, Q_0 is the $N(0, I_n)$ distribution on R^n, G is the group given in the example and $\mathscr{P}$ is just the family $\{N(\mu e_n, \sigma^2 I) | \mu \in R^1, \sigma^2 > 0\}$. Here is a slightly more complicated linear model example.

Example 3.3. On R^n, fix a distribution Q_0 which we think of as a *standardized error distribution* for a linear model [e.g., $Q_0 = N(0, I_n)$]. Let M be a linear subspace of R^n which is regarded as the *regression subspace* of the linear model. Elements of a group G are pairs (a, b) with $a \neq 0$, $a \in R^1$ and $b \in M$. The group operation is

$$(a_1, b_1)(a_2, b_2) = (a_1 a_2, a_1 b_2 + b_1)$$

and the action on R^n is

$$x \to ax + b.$$

The G-invariant statistical model is

$$\mathscr{P} = \{gQ_0 | g \in G\}.$$

To describe this model in more standard terminology, let ε_0 have distribution Q_0 so $\mathscr{L}(\varepsilon_0) = Q_0$. For $(a, b) \in G$,

$$\mathscr{L}((a, b)\varepsilon_0) = \mathscr{L}(b + a\varepsilon_0).$$

Hence an observation from this model can be written

$$Y = b + a\varepsilon_0,$$

where $b \in M$. Assuming ε_0 has mean 0, the mean of Y is $b \in M$ and the covariance matrix of Y is

$$a^2 \operatorname{Cov}(\varepsilon_0),$$

where $\operatorname{Cov}(\varepsilon_0)$ is the covariance matrix of ε_0. Setting $\varepsilon = a\varepsilon_0$ and $\mu = b$, the model for Y is

$$Y = \mu + \varepsilon$$

which is the standard "Y equals mean vector plus error" model common in linear regression. When $\operatorname{Cov}(\varepsilon_0) = I_n$, we are in the case when $\operatorname{Cov}(Y) = \sigma^2 I_n$ for some $\sigma^2 > 0$. Thus, the usual simple regression models are group generated models when it is assumed that the error distribution is some scaled version of a fixed distribution on R^n. □

In many situations, invariant statistical models $\mathscr{P}$ are parametric statistical models having parametric density functions with respect to a fixed σ-finite measure. The proper context to discuss the expression of the invariance in terms of the densities is the following. Consider a topological group G acting measurably on $(\mathbf{X}, \mathscr{B})$ and assume that μ is a σ-finite measure on $(\mathbf{X}, \mathscr{B})$ which is relatively invariant with multiplier χ on G. That is,

$$\int f(g^{-1}x)\mu(dx) = \chi(g)\int f(x)\mu(dx)$$

for $g \in G$ and μ-integrable f.

THEOREM 3.1. *Assume the group G acts on the parameter space Θ and that $\{p(\cdot|\theta)|\theta \in \Theta\}$ is a family of densities with respect to μ. If the densities satisfy*

$$p(x|\theta) = p(gx|g\theta)\chi(g), \tag{3.1}$$

then the parametric family of probability measures $\mathscr{P} = \{P_\theta|\theta \in \Theta\}$ defined by the densities is G-invariant. Further, $gP_\theta = P_{g\theta}$.

PROOF. For each $\theta \in \Theta$ and $g \in G$, it suffices to verify that

$$gP_\theta = P_{g\theta}$$

since this implies $\mathscr{P}$ is G-invariant. For $B \in \mathscr{B}$,

$$\begin{aligned}(gP_\theta)(B) = P_\theta(g^{-1}B) &= \int I_B(gx)p(x|\theta)\mu(dx)\\ &= \chi(g)\int I_B(gx)p(gx|g\theta)\mu(dx)\\ &= \chi(g)\chi(g^{-1})\int I_B(x)p(x|g\theta)\mu(dx)\\ &= P_{g\theta}(B).\end{aligned}$$

□

A converse to Theorem 3.1 is "almost true." That is, consider a parametric family $\mathscr{P} = \{P_\theta|\theta \in \Theta\}$ which is G-invariant and satisfies

$$gP_\theta = P_{g\theta}, \qquad g \in G, \theta \in \Theta.$$

If each P_θ has a density $p(x|\theta)$ with respect to a σ-finite measure μ which is relatively invariant with multiplier χ, then the argument used in Theorem 3.1 shows that (3.1) holds a.e. for each $g \in G$ and $\theta \in \Theta$. Unfortunately, the null set where (3.1) does not hold can depend on both g and θ. However, in all of the interesting cases that I know, a version of the density exists so that (3.1) holds for all x, θ and g. When (3.1) holds for the density $p(\cdot|\theta)$, we say that the family of densities is *invariant* (the multiplier χ is understood to be given by the context), although a better word might be χ-invariant.

EXAMPLE 3.4. Consider a random vector $X \in R^p$ which is multivariate normal with mean vector $\mu \in R^p$ and positive definite covariance $\Sigma = \operatorname{Cov}(X) \in S_p^+$. The density of X with respect to Lebesgue measure dx is

$$p(x|\mu, \Sigma) = \frac{|\Sigma|^{-1/2}}{(\sqrt{2\pi})^p}\exp\left[-\tfrac{1}{2}(x-\mu)'\Sigma^{-1}(x-\mu)\right]. \tag{3.2}$$

Thus, the sample space is R^p and the parameter space is $\Theta = R^p \times S_p^+$. The affine group Al_p acts on R^p by

$$x \to gx + a,$$

where $(g, a) \in \mathrm{Al}_p$ with $g \in \mathrm{Gl}_p$ and $a \in R^p$. When $\mathscr{L}(X) = N(\mu, \Sigma)$, then

$$\mathscr{L}((g, a)X) = N(g\mu + a, g\Sigma g'),$$

so the appropriate action on θ is

$$(\mu, \Sigma) \to (g\mu + a, g\Sigma g').$$

An easy calculation shows that dx is relatively invariant with multiplier

$$\chi(g, a) = |\det(g)|, \qquad (g, a) \in \mathrm{Al}_p.$$

The direct verification of (3.1) entails showing that

$$p(x|\mu, \Sigma) = p(gx + a|g\mu + a, g\Sigma g')|\det(g)|,$$

which is routine.

The parametric family of this example is

$$\mathscr{P} = \left\{N(\mu, \Sigma)|\mu \in R^p, \Sigma \in S_p^+\right\}.$$

With $Q_0 = N(0, I_p)$, it is clear that Al_p acting on Q_0 generates the family $\mathscr{P}$, that is,

$$\mathscr{P} = \left\{(g, a)Q_0|(g, a) \in \mathrm{Al}_p\right\}.$$

Notice that Al_p does not give a one-to-one indexing for this parametric family. There is nothing special about the normal distribution in the example above. Given any fixed density f_1 (with respect to dx) on R^p, let Q_1 be the probability measure defined by f_1. Then the parametric family

$$\mathscr{P}_1 = \left\{(g, a)Q_1|(g, a) \in \mathrm{Al}_p\right\}$$

is generated by Al_p acting on Q_1. A direct calculation shows that $(g, a)Q$ has a density

$$p(x|(g, a)) = f_1(g^{-1}(x - a))|\det(g)|^{-1},$$

which clearly satisfies (3.1). Again, the group Al_p ordinarily does not provide a one-to-one indexing of the parametric family $\mathscr{P}_1$. That is, it is usually the case that

$$(g_1, a_1)Q_1 = (g_2, a_2)Q_1$$

does not imply that $(g_1, a_1) = (g_2, a_2)$.

In the case of the normal distribution, the mean and covariance provided a one-to-one indexing of $\mathscr{P}$. However, this is not the case in general, but when the density f_1 has the form $f_1(x) = k_1(\|x\|^2)$, then the distribution $(g, a)Q_1$ depends on (g, a) only through gg' and a, as in the normal case. □

3.2. Invariant testing problems. In this section, invariant testing problems are described and invariant tests are discussed. The setting for this discussion is a statistical model $\mathscr{P}$ which is invariant under a group G acting on a sample space $(\mathbf{X}, \mathscr{B})$. Thus, the observed random variable X satisfies $\mathscr{L}(X) \in \mathscr{P}$. Consider a testing problem in which the null hypothesis H_0 is that a submodel $\mathscr{P}_0$ of $\mathscr{P}$ actually obtains. In other words, the null hypothesis is that $\mathscr{L}(X) \in \mathscr{P}_0 \in \mathscr{P}$ as opposed to the alternative that $\mathscr{L}(X) \in \mathscr{P} - \mathscr{P}_0$.

DEFINITION 3.2. The above hypothesis testing problem is *invariant under* G if both $\mathscr{P}_0$ and $\mathscr{P}$ are G invariant.

When the testing problem is invariant under G, then it is clear that the alternative $\mathscr{P} - \mathscr{P}_0$ is also invariant under G.

Following standard terminology [e.g., Lehmann (1986)], a test function ϕ is a measurable function from $(\mathbf{X}, \mathscr{B})$ to $[0,1]$ and $\phi(x)$ is interpreted as the conditional probability of rejecting H_0 when the observation is x. The behavior of a test function ϕ is ordinarily described in terms of the *power function*

$$\beta(P) = \mathbf{E}_P \phi(X), \qquad P \in \mathscr{P}.$$

Ideally, one would like to choose ϕ to make $\beta(P) = 0$ for $P \in \mathscr{P}_0$ and $\beta(P) = 1$ for $P \in \mathscr{P} - \mathscr{P}_0$.

When a hypothesis testing problem is invariant under a group G, it is common to see the following "soft" argument to support the use of an *invariant test function*, that is, a test function ϕ which satisfies $\phi(x) = \phi(gx)$ for $x \in \mathbf{X}$ and $g \in G$. This argument is:

> Consider $x \in \mathbf{X}$ and suppose $X = x$ supports H_0. Then we tend to believe $\mathscr{L}(X) \in \mathscr{P}_0$. If we had observed $X = gx$ instead, then $x = g^{-1}X$ and $\mathscr{L}(g^{-1}X) \in \mathscr{P}_0$ when $\mathscr{L}(X) \in \mathscr{P}_0$. Hence we should also believe H_0 if gx obtains. In other words, x and gx should carry the same weight of evidence for H_0. Exactly the same argument holds if x supports H_1.

We now turn to a discussion of two ways of obtaining an invariant test in the special case that $\mathscr{P}$ is a parametric family $\mathscr{P} = \{P_\theta | \theta \in \Theta\}$, G acts on Θ and there is a density $p(\cdot|\theta)$ for P_θ with respect to a σ-finite measure μ. It is assumed further that μ is relatively invariant with multiplier χ and the density $p(\cdot|\theta)$ satisfies Equation (3.1), i.e.,

$$p(x|\theta) = p(gx|g\theta)\chi(g).$$

In this notation, the invariance of the hypothesis testing problem means that

$$\mathscr{P}_0 = \{P_\theta | \theta \in \Theta_0\}$$

is G invariant so the set $\Theta_0 \in \Theta$ is G invariant. The likelihood ratio statistic for testing H_0: $\theta \in \Theta_0$ versus H_1: $\theta \in \Theta - \Theta_0 = \Theta_1$ is ordinarily defined by

$$\Lambda(x) = \frac{\sup_{\theta \in \Theta_0} p(x|\theta)}{\sup_{\theta \in \Theta} p(x|\theta)}.$$

This statistic is then used to define a test function ϕ via

$$\phi(x) = \begin{cases} 1, & \text{if } \Lambda(x) < c, \\ 0, & \text{if } \Lambda(x) \geq c, \end{cases}$$

where c is some appropriately chosen constant. Because of our assumptions on $p(\cdot|\theta)$ and the invariance of Θ_0 and Θ, it is readily shown that $\Lambda(x) = \Lambda(gx)$ and hence that $\phi(x) = \phi(gx)$. In other words, the test defined by the likelihood ratio statistic is invariant under any group for which the testing problem is invariant.

A second method which can sometimes be employed to define an invariant test involves the use of relatively invariant measures defined on Θ_0 and Θ_1. More precisely, assume we can find measures ξ_0 and ξ_1 on Θ_0 and Θ_1, respectively, which are relatively invariant with the same multiplier χ_1. Assuming the following expression makes sense, let

$$T(x) = \frac{\int_{\Theta_0} p(x|\theta)\xi_0(d\theta)}{\int_{\Theta_1} p(x|\theta)\xi_1(d\theta)}.$$

Then T can be used to define a test function ϕ via

$$\phi(x) = \begin{cases} 1, & \text{if } T(x) < c, \\ 0, & \text{if } T(x) \geq c, \end{cases}$$

where c is the omnipresent constant. The assumed invariance of the density $p(\cdot|\theta)$ and the relative invariance of ξ_0 and ξ_1 combine to imply that $T(x) = T(gx)$ and hence that the test ϕ is an invariant test.

The main reason for introducing the invariant tests described above is to raise some questions for which answers are provided in later lectures—in particular, how to select ξ_0 and ξ_1. Under some rather restrictive conditions, it is shown in a later lecture how to choose the measures ξ_0 and ξ_1 so that the test defined by the statistic T is a "good" invariant test. It is also shown that the likelihood ratio test does not necessarily yield a "good" invariant test when one exists.

3.3. Equivariant estimators. As in the last section, consider a group G which acts on a sample space $(\mathbf{X}, \mathscr{B})$ and a parameter space Θ in such a way that the parametric model $\mathscr{P} = \{P_\theta | \theta \in \Theta\}$ is invariant. A density $p(x|\theta)$ is assumed to exist and the invariance condition (3.1) is to hold throughout this section. Thus, the dominating measure μ is relatively invariant with multiplier χ. In this context, a point estimator t mapping $\mathbf{X}$ to Θ is *equivariant* if

$$t(gx) = gt(x). \tag{3.3}$$

A soft argument leading to the consideration of equivariant estimators is:

> Consider t_0 as an estimator of θ. When $\mathscr{L}(X) = P_\theta$, then $t_0(X)$ is supposed to estimate θ. But, when $\mathscr{L}(X) = P_\theta$, $\mathscr{L}(gX) = P_{g\theta}$ so $t_0(gX)$ should estimate $g\theta$. However, $t_0(X)$ estimates θ, so $gt_0(X)$ estimates $g\theta$. Equating these two estimators of θ leads to $t_0(gX) = gt_0(X)$ and hence estimators satisfying (3.3).

The method of maximum likelihood leads to an equivariant estimator when the maximum likelihood estimator is unique. This result is a consequence of the following constructive method for finding a maximum likelihood estimator.

Theorem 3.2. *Consider X with G-invariant density $p(\cdot|\theta)$ and fix $x_0 \in \mathbf{X}$. Assume that $\theta_0 \in \Theta$ uniquely maximizes $p(x_0|\theta)$ as θ varies over Θ, so $p(x_0|\theta) \le p(x_0|\theta_0)$ with equality iff $\theta = \theta_0$. For $x \in O_{x_0}$, the orbit of x_0, write $x = g_x x_0$ for some $g_x \in G$ and set*

$$\hat{\theta}(x) = g_x \theta_0.$$

Then for $x \in O_{x_0}$, $\hat{\theta}$ is the maximum likelihood estimator of θ, is unique and is equivariant.

Proof. The proof is not hard and can be found in Eaton [1983, page 259–260]. □

The import of Theorem 3.2 is that, in invariant situations, the maximum likelihood estimators can be found by simply selecting some convenient point x_0 from each orbit in $\mathbf{X}$ and then calculating the maximum likelihood estimator, say θ_0, for that x_0. The value of $\hat{\theta}(x)$ for other x's in the same orbit is calculated by finding a g_x such that $g_x x_0 = x$ and setting $\hat{\theta}(x) = g_x \theta_0$. This orbit-by-orbit method of solution arises in other contexts later.

The results of Theorem 3.2 are valid for other methods of estimation also. For example, certain nonparametric methods can be characterized as choosing an estimator $t_0(x)$ so as to maximize a function

$$H(x|\theta) \ge 0$$

as θ ranges over Θ. If the function H satisfies

$$H(x|\theta) = H(gx|g\theta)\chi_0(g)$$

for some multiplier χ_0, then Theorem 2.3 applies [just replace the density $p(x|\theta)$ by $H(x|\theta)$ in the statement of Theorem 3.2]. In other words, the orbit-by-orbit method applies and the resulting estimator is equivariant and unique.

In a Bayesian context, inferential statements about θ are in the form of probability distributions on Θ which depend on x. These probability distributions are often obtained from a measure ξ on Θ. The measure ξ is a *prior distribution* if $0 < \xi(\Theta) < +\infty$ and is an *improper prior distribution* if $\xi(\Theta) = +\infty$. Given ξ, let

$$m(x) = \int p(x|\theta)\xi(d\theta)$$

and assume that $0 < m(x) < +\infty$ for all $x \in \mathbf{X}$. Then define

$$q(\theta|x) = \frac{p(x|\theta)}{m(x)}$$

so that

$$Q(B|x) = \int I_B(\theta) q(\theta|x)\xi(d\theta)$$

determines a probability measure $Q(\cdot|x)$ on Θ. For B fixed $Q(B|\cdot)$ is a measurable function defined on $(\mathbf{X}, B)$. Thus Q is a Markov kernel (randomized decision rule, posterior distribution, etc.) as discussed in Example 2.19. In this context, the appropriate notion of invariance is that defined in Example 2.19, namely Q is invariant if $gQ = Q$. That is, if

$$Q(B|x) = Q(g^{-1}B|g^{-1}x) \tag{3.4}$$

for measurable sets $B \subset \Theta$, $x \in \mathbf{X}$ and $g \in G$. Here is a condition which implies that (3.4) holds:

THEOREM 3.3. *Assume the measure ξ on Θ is relatively invariant with some multiplier χ_1. Then Q satisfies* (3.4).

PROOF. For $g \in G$,

$$\begin{aligned} Q(g^{-1}B|g^{-1}x) &= \int I_B(g\theta)q(\theta|g^{-1}x)\xi(d\theta) \\ &= \frac{\int I_B(g\theta)p(g^{-1}x|\theta)\xi(d\theta)}{\int p(g^{-1}x|\theta)\xi(d\theta)}. \end{aligned}$$

Using (3.1), we have

$$Q(g^{-1}B|g^{-1}x) = \frac{\int I_B(g\theta)p(x|g\theta)\xi(d\theta)}{\int p(x|g\theta)\xi(d\theta)}.$$

The assumed invariance of ξ now yields (3.4). $\square$

Again, the question of how to choose ξ from the class of all relatively invariant measures naturally arises. This is discussed in later lectures.

Finally, the relationship between the equivariance of a point estimator and (3.4) requires a comment. Given any point estimator $t_0(x)$, the natural way to identify t_0 with a Markov kernel Q_0 is to let $Q_0(\cdot|x)$ be degenerate at the point $t_0(x) \in \Theta$, that is,

$$Q_0(B|x) = \begin{cases} 1, & \text{if } t_0(x) \in B, \\ 0, & \text{otherwise.} \end{cases}$$

With this identification, it is routine to show that (3.4) holds iff t_0 is equivariant.

Here is a simple example. More interesting and complicated examples appear in later lectures.

EXAMPLE 3.5. Suppose $X \in R^1$ is $N(\theta, 1)$, so $\Theta = R^1$. Obviously the model $\mathscr{P} = \{N(\theta,1)|\theta \in R^1\}$ is invariant under the group $G = R^1$ acting on $\mathbf{X}$ and Θ via translation. For this example t is an equivariant point estimator iff

$$t(x+g) = t(x) + g$$

for all $x, g \in R^1$. Setting $g = -x$, we see that t is equivariant iff

$$t(x) = x + a,$$

where a is some fixed real number. The choice $a = 0$ gives the maximum likelihood estimator. To generate invariant distributions on Θ, consider the relatively invariant measure

$$\xi(d\theta) = e^{\theta b}\, d\theta,$$

where b is a fixed real number. (These are all the relatively invariant Radon measures on R^1 up to positive multiples.) For x given, this ξ gives

$$Q(\cdot|x) = N(x+b,1) \quad \text{on } \Theta = R^1.$$

In words, the formal posterior distribution is normal with mean $x + b$ and variance 1. Of course, the "usual" choice is $b = 0$ in which case $\xi(d\theta) = d\theta$ is the Haar measure on $G = \Theta$. In more complicated examples where G is not unimodular, the choice is not so obvious. □

3.4. Linear models. In order to motivate our discussion of linear models, first consider what is commonly called the multivariate analysis of variance model (the MANOVA model)

$$Y = XB + E. \tag{3.5}$$

Here, X: $n \times k$ is a known matrix of rank k, B: $k \times p$ is a matrix of unknown regression parameters and E: $n \times p$ is a matrix of random variables (errors). A standard assumption concerning E is that the rows of E are iid multivariate normal with mean 0 and common positive definite covariance, say C: $p \times p$. This assumption on E is often written

$$\mathscr{L}(E) = N(0, I_n \otimes C), \tag{3.6}$$

where $I_n \otimes C$ denotes the Kronecker product of the $n \times n$ identity matrix I_n and C. Thus the $n \times p$ error matrix has a normal distribution (with mean 0 and the specified covariance) on the vector space of $n \times p$ real matrices $\mathscr{L}_{p,n}$. The standard coordinate inner product on $\mathscr{L}_{p,n}$ is

$$\langle x, y\rangle = \operatorname{tr} xy' = \sum_i \sum_j x_{ij} y_{ij},$$

where tr denotes the trace. That E has covariance $I_n \otimes C$ means that

$$\operatorname{cov}\{\langle x, E\rangle, \langle y, E\rangle\} = \langle x, (I_n \otimes C)y\rangle,$$

where cov denotes ordinary covariance between real random variables. As usual, the Kronecker product $I_n \otimes C$ is the linear transformation on $\mathscr{L}_{p,n}$ to $\mathscr{L}_{p,n}$ defined by

$$(I_n \otimes C)x = I_n x C' = xC'.$$

An alternative way to write (3.5) is

$$Y = \mu + E, \tag{3.7}$$

where E is the error vector as before and $\mu = XB$ is the mean vector for Y. Thus, the space of possible values for μ is the linear subspace of $\mathscr{L}_{p,n}$,

$$M = \{\mu | \mu = XB, B\colon k \times p\} \subset \mathscr{L}_{p,n}. \tag{3.8}$$

When the distributional assumption (3.6) holds, then

$$\mathscr{L}(Y) = N(\mu, I_n \otimes C), \tag{3.9}$$

where $\mu \in M$ and C is some $p \times p$ positive definite covariance matrix. Thus the parametric model for $\mathscr{L}(Y)$ is

$$\mathscr{P} = \{N(\mu, I_n \otimes C) | \mu \in M, C \in S_p^+\}.$$

To describe the invariance of this model, let $M_0 \subset R^n$ be the linear subspace spanned by the columns of the matrix X and set

$$G_0 = \{g | g \in O_n, g(M_0) \subset M_0\}.$$

Elements of G_0 are orthogonal transformations on R^n to R^n which have M_0 and hence the orthogonal complement $M_0^{\perp}$, as invariant subspaces. Note that if $\mu \in M$, then the matrix product $g\mu$ is also in M for $g \in G_0$, because the columns of μ are elements of M_0.

Now, let G have elements which are triples (g, a, α) with $g \in G_0$, $a \in \mathrm{Gl}_p$ and $\alpha \in M$. The action of (g, a, α) on $\mathscr{L}_{p,n}$ is

$$(g, a, \alpha)x = gxa' + \alpha$$

and the group composition is

$$(g_1, a_1, \alpha_1)(g_2, a_2, \alpha_2) = (g_1 g_2, a_1 a_2, g_1 \alpha_2 a_1' + \alpha_1).$$

When (3.9) holds, then

$$\mathscr{L}((g, a, \alpha)Y) = N(g\mu a' + \alpha, I_n \otimes (aCa')),$$

which is again in $\mathscr{P}$. Thus the MANOVA model $\mathscr{P}$ is invariant under G. It is well known that the maximum likelihood estimator for μ in this model is

$$\hat{\mu} = P_0 Y,$$

where P_0: $n \times n$ is the orthogonal projection onto $M_0 \subset R^n$. Further, $\hat{\mu}$ is the unique unbiased estimator of μ based on the sufficient statistic for $\mathscr{P}$ and $\hat{\mu}$ is the best linear unbiased estimator of μ. The invariance of the model $\mathscr{P}$ implies that $\hat{\mu}$ is an equivariant estimator of μ where the action of G on M is

$$\mu \to g\mu a' + \alpha.$$

It should be mentioned that the linear transformation on $\mathscr{L}_{p,n}$ to $\mathscr{L}_{p,n}$ defined by

$$x \to P_0 x$$

is just the orthogonal projection onto M in the inner product space $(\mathscr{L}_{p,n}, \langle \cdot , \cdot \rangle)$.

The aspect of the MANOVA model with which the rest of this section deals is the equivariance of the estimator of the mean vector. For this discussion, we first describe the Gauss–Markov theorem for the so-called *regular linear models*. Consider finite dimensional inner product space $(V, (\cdot, \cdot))$. By a linear model for a random vector Y with values in V, we mean a model of the form

$$Y = \mu + \varepsilon, \tag{3.10}$$

where

(i) the random vector ε has mean 0 and covariance $\Sigma = \mathrm{Cov}(\varepsilon)$ assumed to lie in some known set γ of positive definite linear transformations;

(ii) the mean vector μ of Y lies in a known subspace M of V which is called the *regression subspace*.

Thus, the pair (M, γ) determines the assumed mean and covariance structure for Y taking values in $(V, (\cdot, \cdot))$. In what follows it is assumed that the identity covariance I is in γ. This assumption is without loss of generality since any pair (M, γ) can be transformed via one element of γ to another pair (M_1, γ_1) with $I \in \gamma_1$.

DEFINITION 3.3. The pair (M, γ) (with $I \in \gamma$) is a *regular linear model* for Y if

$$\Sigma(M) \subset M, \qquad \Sigma \in \gamma. \tag{3.11}$$

In other words, the model is regular if the regression subspace is an invariant subspace under each possible covariance $\Sigma \in \gamma$. That the MANOVA model is regular is readily verified. The condition (3.11) is equivalent to the condition

$$\Sigma A_0 = A_0 \Sigma, \qquad \Sigma \in \gamma, \tag{3.12}$$

where A_0 is the orthogonal projection onto the subspace M.

Consider any linear unbiased estimator AY of $\mu \in M$, that is, A is a linear transformation from V to V which satisfies

$$Ax = x, \qquad x \in M. \tag{3.13}$$

Here is one version of the Gauss–Markov theorem which compares linear unbiased estimators in terms of covariance:

THEOREM 3.4 (Gauss–Markov theorem). *Assume the pair (M, γ) determines a regular linear model for Y and let A_0 be the orthogonal projection onto M. Then, for any linear unbiased estimator AY of $\mu \in M$,*

$$\operatorname{Cov}(A_0 Y) \le \operatorname{Cov}(AY),$$

where $\le$ is in the sense of nonnegative definiteness. That is, $Cov(AY) - Cov(A_0 Y)$ is nonnegative definite.

PROOF. With $\Sigma = \operatorname{Cov}(Y)$,

$$\begin{aligned}\operatorname{Cov}(AY) &= A\Sigma A' \\ &= A_0\Sigma A_0 + (A - A_0)\Sigma A_0 \\ &\quad + A_0 \Sigma (A - A_0)' + (A - A_0)\Sigma(A - A_0)'.\end{aligned}$$

However, the term $(A - A_0)\Sigma A_0$ is 0. To see this, first notice that $\Sigma A_0 = A_0 \Sigma$ due to the regularity of the linear model. But (3.13) implies that $(A - A_0)A_0 = 0$, so

$$(A - A_0)\Sigma A_0 = (A - A_0)A_0\Sigma = 0.$$

Thus $A_0\Sigma(A - A_0)'$ is also 0 as it is the transpose of $(A - A_0)\Sigma A_0$. Hence

$$\begin{aligned}\operatorname{Cov}(AY) &= A_0\Sigma A_0 + (A - A_0)\Sigma(A - A_0)' \\ &= \operatorname{Cov}(A_0 Y) + (A - A_0)\Sigma(A - A_0)'.\end{aligned}$$

□

Now consider a linear model

$$Y = \mu + \varepsilon,$$

where again $\mu \in M$, the error vector ε has mean zero and satisfies the following invariance assumption:

$$\mathcal{L}(\varepsilon) = \mathcal{L}(g_0\varepsilon),$$

where

$$g_0 = I - 2A_0.$$

As usual, A_0 is the orthogonal projection onto M, so $g_0^2 = I$ and $g_0^{-1} = g_0$. This model for Y is invariant under the transformations

$$x \to gx + a, \qquad x \in V,$$

where g is either g_0 or I and $a \in M$. In other words when $Y = \mu + \varepsilon$,

$$gY + a = (g\mu + a) + g\varepsilon = \mu^* + \varepsilon^*,$$

where $\mu^* \in M$ and ε^* has the same distribution as ε. Hence the mean of $gY + a$ is still in M and the error vector has the same distribution for $gY + a$ as for Y. The group in question, say G, has elements which are pairs (g, a) with $a \in M$ and g is g_0 or I. The group operation is

$$(g_1, a_1)(g_2, a_2) = (g_1g_2, g_1a_2 + a_1).$$

The appropriate equivariance of a point estimator t of μ is

$$t(gx + a) = gt(x) + a \tag{3.14}$$

because μ is mapped into $g\mu + a$ by the group element (g, a). The following result shows that an equivariant estimator of μ is just the Gauss–Markov estimator A_0Y of Theorem 3.4.

THEOREM 3.5. *Suppose t: $V \to M$ satisfies* (3.14) *for* $(g, a) \in G$. *Then* $t(x) = A_0x$ *for* $x \in V$.

PROOF. For $x \in V$, write $x = x_1 + x_2$ where $x_1 \in M$ and $x_2 \in M^{\perp}$. Here, $M^{\perp}$ is the orthogonal complement of M. Choosing $g = I$ and $a = -x_1$ in (3.14) yields

$$t(x_1 + x_2 - x_1) = t(x_1 + x_2) - x_1$$

so that

$$t(x_1 + x_2) = x_1 + t(x_2).$$

Now, in (3.14) pick $a = 0$, $x = x_2 \in M^{\perp}$ and $g = g_0$. From (3.14) we have

$$t(g_0x_2) = g_0t(x_2).$$

But $g_0x_2 = x_2$ as $x_2 \in M^{\perp}$, and $g_0t(x_2) = -t(x_2)$ because t takes values in M. Thus,

$$t(x_2) = -t(x_2)$$

so $t(x_2) = 0$ for all $x_2 \in M^{\perp}$. Therefore

$$t(x) = t(x_1 + x_2) = x_1 + t(x_2) = A_0x. \qquad \square$$

The invariance argument used to characterize the equivariant estimator A_0Y depends on the assumption

$$\mathcal{L}(\varepsilon) = \mathcal{L}(g_0\varepsilon),$$

where $g_0 = I - 2A_0$. Hence, if ε has a covariance, say $\Sigma = \text{Cov}(\varepsilon)$, this invariance assumption on $\mathcal{L}(\varepsilon)$ implies that

$$\Sigma = g_0 \Sigma g_0. \tag{3.15}$$

However, this condition is exactly the same as the regularity assumption which led to Theorem 3.2. To see this, (3.15) is

$$\begin{aligned}\Sigma &= (I - 2A_0)\Sigma(I - 2A_0\Sigma)\\ &= \Sigma - 2A_0\Sigma - 2\Sigma A_0 + 4A_0\Sigma A_0,\end{aligned}$$

which yields

$$2A_0\Sigma A_0 = A_0\Sigma + \Sigma A_0.$$

Now, multiplying this on the left by A_0 gives

$$A_0\Sigma A_0 = A_0\Sigma$$

while multiplication on the right gives

$$A_0\Sigma A_0 = \Sigma A_0$$

so that $\Sigma A_0 = A_0\Sigma$. This is just (3.12) which is equivalent to (3.11). Thus, the invariance assumption on $\mathcal{L}(\varepsilon)$ is very closely connected with the assumption on γ in the definition of regularity.

CHAPTER 4

Models Invariant under Compact Groups

Our goal here is to understand the structure of probability measures which are invariant under a compact group. In the first section, a basic representation theorem is proved and is interpreted in terms of random variables. Section 2 contains some basic examples while applications to some robustness problems are given in Section 3.

4.1. A representation theorem. Throughout this section, G is a compact group and ν is the unique left (and right) invariant probability measure on G. In some situations, it is convenient to express certain equations in terms of a "random group element" U which has distribution ν. This is written $\mathscr{L}(U) = \nu$ and we say U has a uniform distribution on G. What this means is that U is a random element of G and the expectation of any bounded measurable function of U is computed as

$$\mathscr{E}f(U) = \int_G f(u)\nu(du).$$

Naturally, the distribution of U is characterized by its invariance. That is, the equation

$$\mathscr{L}(U) = \mathscr{L}(gU), \qquad g \in G,$$

characterizes the distribution of U because of the uniqueness of the invariant measure ν.

EXAMPLE 4.1. Let G be the group of $n \times n$ real orthogonal matrices O_n. The existence of ν on O_n is given by general theory, but here we outline a "construction" of U, and hence ν, which uses the normal distribution. Let $X_1, \ldots, X_n$ be iid $N(0, I_n)$ random vectors in R^n. That $X_1, \ldots, X_n$ are linearly independent with probability 1 is not hard to prove [see Proposition 7.1 in Eaton (1983) for example]. Let $Y_1,\dots,Y_n$ be an orthonormal basis for R^n obtained by performing the Gram–Schmidt orthogonalization procedure to $X_1,\dots,X_n$ in that order. Consider $U \in O_n$ with columns $Y_1,\dots,Y_n$ so that $U = U(X_1,\dots,X_n)$ is a function of $X_1,\dots,X_n$. Hence U is a random element of O_n.

Using the definition of the Gram–Schmidt procedure, it is easy to show

$$U(gX_1,\dots,gX_n) = gU(X_1,\dots,X_n) \tag{4.1}$$

for each $g \in O_n$. In other words, the orthonormal basis one obtains from $gX_1,\dots,gX_n$ is the same as one obtains by first constructing an orthonormal basis from $X_1,\dots,X_n$ and then transforming the basis by $g \in O_n$.

The claim is that U is uniform on O_n. To see this, first observe that

$$\mathscr{L}(X_1,\dots,X_n) = \mathscr{L}(gX_1,\dots,gX_n)$$

because $X_1,\dots,X_n$ are iid $N(0, I_n)$. Using (4.1), we then have

$$\begin{aligned}\mathscr{L}(U) &= \mathscr{L}(U(X_1,\dots,X_n))\\ &= \mathscr{L}(U(gX_1,\dots,gX_n)) = \mathscr{L}(gU(X_1,\dots,X_n))\\ &= \mathscr{L}(gU).\end{aligned}$$

But, as remarked earlier, the relation $\mathscr{L}(U) = \mathscr{L}(gU)$ for $g \in O_n$ characterizes the distribution of U as being uniform. With $\nu = \mathscr{L}(U)$, ν is the unique invariant probability on O_n. □

Now consider a space $\mathbf{X}$ and suppose the compact group G acts topologically on $\mathbf{X}$. Let P be a probability measure defined on the Borel σ-algebra of $\mathbf{X}$ and define a new probability measure P_1 by

$$P_1 = \int_G gP\nu(dg). \tag{4.2}$$

Equation (4.2) means

$$P_1(B) = \int_G (gP)(B)\nu(dg) = \int_G P(g^{-1}B)\nu(dg), \tag{4.3}$$

or in terms of $f \in K(\mathbf{X})$,

$$\int f(x)P_1(dx) = \int_G\int_{\mathbf{X}} f(gx)P(dx)\nu(dg). \tag{4.4}$$

Using (4.3), it is obvious that $hP_1 = P_1$ because ν is the invariant probability measure on G. Thus, averaging gP with respect to ν produces a G invariant probability. Also observe that if P in (4.2) is invariant, then $P = P_1$.

There is an alternative way to write (4.2) which is also interesting. To this end, let U be uniform on G and for $x \in \mathbf{X}$, consider the random element $Ux \in \mathbf{X}$. For a Borel set $B \subset \mathbf{X}$, the probability that Ux is in B is

$$\text{Prob}(Ux \in B) = \nu\{g\colon gx \in B\}.$$

The induced distribution of Ux on $\mathbf{X}$ is denoted by μ_x. In other words,

$$\mu_x(B) = \nu\{g|gx \in B\},$$

so that

$$\mathscr{E}f(Ux) = \int_{\mathscr{X}} f(z)\mu_x(dz) = \int_G f(gx)\nu(dg). \tag{4.5}$$

THEOREM 4.1. *For each $x \in \mathbf{X}$, the probability μ_x is G-invariant.*

PROOF. Since $\mathscr{L}(Ux) = \mu_x$,

$$g\mu_x = \mathscr{L}(g(Ux)) = \mathscr{L}((gU)x) = \mathscr{L}(Ux) = \mu_x,$$

where the next to the last equality follows from the uniformity of U on G. □

The random variable $Ux \in \mathbf{X}$ obviously takes its value in the orbit of x since $U \in G$. Thus $\mu_x = \mathscr{L}(Ux)$ is a G-invariant probability measure on $O_x = \{gx | g \in G\}$. Since G acts transitively on O_x and G is compact, Theorem 2.2 shows that μ_x is in fact the unique invariant probability measure on O_x (modulo checking the regularity conditions needed to apply Theorem 2.2). This turns out to be a useful way to think about μ_x.

Since μ_x is G-invariant for each $x \in \mathbf{X}$, the average of μ_x (over $\mathbf{X}$) is also G-invariant. In symbols,

$$P_1 = \int \mu_x P(dx)$$

is invariant. This equation means

$$P_1(B) \equiv \int \mu_x(B) P(dx)$$

for each Borel set B. The next result shows that every invariant probability has such a representation as an average of the μ_x's.

THEOREM 4.2. *Suppose P is a G-invariant probability measure on $\mathbf{X}$. Then*

$$P = \int_{\mathbf{X}} \mu_x P(dx). \tag{4.6}$$

PROOF. When P is invariant, Equation (4.2) is

$$P = \int gP\nu(dg),$$

which expressed as in (4.4), with $f = I_B$, is

$$P(B) = \int_G \int_{\mathbf{X}} I_B(gx) P(dx)\nu(dg).$$

Interchanging integrals and using the definition of μ_x yields

$$\begin{aligned} P(B) &= \int_{\mathbf{X}} \int_G I_B(gx)\nu(dg) P(dx) \\ &= \int_{\mathbf{X}} \mu_x(B) P(dx) \end{aligned}$$

which is just (4.6). □

Theorem 4.2 is unsatisfactory because of the following. Notice that

$$\mu_{gx} = \mathscr{L}(Ugx) = \mathscr{L}(Ux) = \mu_x$$

because $\nu = \mathscr{L}(U)$ is both left- and right-invariant. Thus, $x \to \mu_x$ is an invariant function and hence can be written as a function of a maximal invariant, say $t(x)$. Thus it should be possible to express the average (4.6) as an average over the maximal invariant (i.e., over the orbits in $\mathbf{X}$) rather than over the whole space $\mathbf{X}$. In order to make this precise, we need the notion of a measurable cross section. The Borel σ-algebra of $\mathbf{X}$ is denoted by $\mathscr{B}$.

DEFINITION 4.1. A subset $\mathbf{Y} \subset \mathbf{X}$ is a *measurable cross section* if:

(i) $\mathbf{Y}$ is measurable.
(ii) For each x, $\mathbf{Y} \cap O_x$ consists of exactly one point, say $y(x)$.
(iii) The function t defined on $\mathbf{X}$ to $\mathbf{Y}$ by $t(x) = y(x)$ is $\mathscr{B}$ measurable when $\mathbf{Y}$ has the σ-algebra $\{B \cap Y | B \in \mathscr{B}\} = \mathscr{B}_1$.

Assume $\mathbf{Y}$ is a measurable cross-section.

THEOREM 4.3. *For each probability Q defined on $(\mathbf{Y}, \mathscr{B}_1)$, the measure*

$$P = \int_{\mathbf{Y}} \mu_y Q(dy) \tag{4.7}$$

is a G-invariant probability on $(\mathbf{X}, \mathscr{B})$. Conversely, if P is a G-invariant probability on $(\mathbf{X}, \mathscr{B})$, then there exists a probability Q on $(\mathbf{Y}, \mathscr{B}_1)$ such that (4.7) holds.

PROOF. Equation (4.7) means that

$$P(B) = \int_{\mathbf{Y}} \mu_y(B) Q(dy)$$

or in terms of a bounded measurable function f,

$$\int f(x) P(dx) = \int_{\mathbf{Y}} \int_G f(gy) \nu(dg) Q(dy). \tag{4.8}$$

Because $(g, x) \to gx$ is jointly continuous and hence jointly measurable, joint measurability of $(g, y) \to f(gy)$ is easily verified. Thus (4.8) makes sense. That (4.7) defines a G-invariant probability is easily checked because μ_y is G-invariant. For the converse, consider a probability P which is invariant. Then, for any bounded measurable function f defined on $\mathbf{X}$, we have

$$\int f(x) P(dx) = \int f(gx) P(dx)$$

for $g \in G$. Integration then yields

$$\int_{\mathbf{X}} f(x) P(dx) = \int_G \int_{\mathbf{X}} f(gx) P(dx) \nu(dg) = \int_{\mathbf{X}} \int_G f(gx) \nu(dg) P(dx). \tag{4.9}$$

For x fixed, there exists a $g_0 \in G$ such that $x = g_0 t(x)$ because x and $t(x)$ are in the same orbit. The invariance of ν gives

$$\int_G f(gx)\nu(dg) = \int_G f(gg_0 t(x))\nu(dg) = \int f(gt(x))\nu(dg).$$

The assumed measurability of t implies that

$$H(t(x)) = \int_G f(gt(x))\nu(dg)$$

is a measurable function. Define the measure Q on $(Y, \mathscr{B}_1)$ by

$$Q(B) = P(t^{-1}(B))$$

so that

$$\int_{\mathbf{X}} H(t(x))P(dx) = \int_{\mathbf{Y}} H(y)Q(dy).$$

Thus from (4.9), we have

$$\int_{\mathbf{X}} f(x)P(dx) = \int_{\mathbf{X}}\int_G f(gt(x))\nu(dg)P(dx),$$

which is just (4.8). □

The interpretation of Theorem 4.3 in terms of random variables is the following. In our earlier notation, let U be uniform on G. Also let Y be a random variable taking values in $(\mathbf{Y}, \mathscr{B}_1)$ which is independent of U. That is, U and Y are defined on some probability space for which they are independent. Now, form

$$X = UY \in \mathbf{X},$$

where by UY, we mean the group element U acting on Y. Thus X is a random variable in $\mathbf{X}$. Because U and Y are independent,

$$\mathscr{L}(gX) = \mathscr{L}(g(UY)) = \mathscr{L}(UY) = \mathscr{L}(X)$$

since

$$\mathscr{L}(gU) = \mathscr{L}(U), \qquad g \in G.$$

With $Q = \mathscr{L}(Y)$ and $P = \mathscr{L}(X)$, it follows immediately that for any bounded measurable function f,

$$\mathscr{E}f(X) = \int_{\mathbf{X}} f(x)P(dx) = \mathscr{E}f(UY) = \int_{\mathbf{Y}}\int_G f(gy)\nu(dg)Q(dy),$$

which is just (4.8) again. Conversely, suppose X takes values in $\mathbf{X}$ and $\mathscr{L}(X) = P = \mathscr{L}(gX)$. Then Equation (4.8) immediately implies the existence of two random variables $U \in G$ and $Y \in \mathbf{Y}$ which are independent, $\mathscr{L}(U) = \nu$ and $\mathscr{L}(Y) = Q$, such that $\mathscr{L}(X) = \mathscr{L}(UY)$. Summarizing this gives:

THEOREM 4.4. *Let U be uniform on G and let* $\mathbf{Y}$ *be a measurable cross section. For a random variable $X \in \mathbf{X}$, the following are equivalent*:

(i) $\mathscr{L}(X) = \mathscr{L}(gX)$, $g \in G$.

(ii) *There exists a random variable $Y \in \mathbf{Y}$ which is independent of U such that* $\mathscr{L}(X) = \mathscr{L}(UY)$.

Various versions of Theorems 4.3 and 4.4 and related results appear in the mathematical and statistical literature. Here are a few relevant references: Wijsman (1957), Farrell (1962), Hall, Wijsman and Ghosh (1965), Dawid (1977), Eaton and Kariya (1984) and the references therein.

4.2. Some standard examples. In the examples below, the random variable notation of Theorem 4.4 is used rather than the more cumbersome notation of Theorem 4.3.

EXAMPLE 4.2. This example is related to exchangeable 0-1 valued random variables. The space $\mathbf{X}$ consists of the set of all n vectors x whose coordinates are only 0 or 1. Thus $\mathbf{X}$ has 2^n elements and we usually write $\mathbf{X} = \{0,1\}^n$. The group $\mathscr{P}_n$ of $n \times n$ permutation matrices acts on the vectors x by matrix multiplication. Of course, if the random vector $X \in \mathbf{X}$ satisfies $\mathscr{L}(X) = \mathscr{L}(gX)$, $g \in \mathscr{P}_n$, then X is *exchangeable*. To apply Theorem 4.4, let $\mathbf{Y} = \{y_0, y_1, \dots, y_n\}$, where $y_i \in \mathbf{X}$ has its first i coordinates equal to 1 and the remaining coordinates equal to 0. That $\mathbf{Y}$ is a cross section is clear.

To say U is uniform on $\mathscr{P}_n$ means that U is picked at random from the set of $n!$ permutation matrices. Theorem 4.4 implies that X is exchangeable iff

$$\mathscr{L}(X) = \mathscr{L}(UY),$$

where Y has an arbitrary distribution on $\mathbf{Y}$ and is independent of U. In other words, X is exchangeable iff X is generated by first picking a $y_i \in \mathbf{Y}$ according to some distribution and then randomly permuting the elements of the picked y_i. □

EXAMPLE 4.3. Here we return to the spherical distributions on $R^n = \mathbf{X}$. With $G = O_n$, a random vector $X \in R^n$ which satisfies $\mathscr{L}(X) = \mathscr{L}(gX)$, $g \in O_n$ has a spherical distribution. Let y_0 be a fixed vector of length 1 in R^n and set

$$\mathbf{Y} = \left\{\alpha y_0 | \alpha \in R^1, \alpha \geq 0\right\}.$$

Since the orbits in R^n are spheres of a given radius, it is clear that $\mathbf{Y}$ intersects each orbit in exactly one point. In other words, an orbit is $\{gx | g \in O_n\}$ and $\|x\| y_0$ is the intersection of $\mathbf{Y}$ and $\{gx | g \in O_n\}$. That $\mathbf{Y}$ is measurable and that $t(x) = \|x\| y_0$ is a measurable function is clear, so $\mathbf{Y}$ is a measurable cross section.

As an Example 4.1, $U \in O_n$ is a random orthogonal matrix. For a random variable $Y \in \mathbf{Y}$, write $Y = Ry_0$ where R is a nonnegative random variable and independent of U. Then X is spherical iff

$$\mathscr{L}(X) = \mathscr{L}(U(Ry_0)) = \mathscr{L}(RUy_0)$$

for some $R \geq 0$ independent of U. Note that Uy_0 has a uniform distribution on $\{x | x \in R^n, \|x\| = 1\}$ so that X is spherical iff X is generated by first picking a radius R and then independently picking a point Uy_0 at random on the surface of the unit sphere. □

EXAMPLE 4.4. This example generalizes the previous one. The space $\mathbf{X}$ of this example consists of the set of all $n \times p$ real matrices of rank p so $p \leq n$. The group $G = O_n$ acts topologically on the left of $\mathbf{X}$ by matrix multiplication, $x \to gx$ for $x \in \mathbf{X}$ and $g \in O_n$. There are two rather natural choices for a cross section in this example. The first is the set

$$\mathbf{Y}_1 = \left\{ \begin{pmatrix} s \\ 0 \end{pmatrix} \in \mathbf{X} \middle| s \text{ is } p \times p \text{ and positive definite} \right\}.$$

This choice stems from the factorization

$$x = g\begin{pmatrix} s \\ 0 \end{pmatrix}, \qquad g \in O_n$$

for each $x \in \mathbf{X}$ where

$$s = (x'x)^{1/2}$$

is the unique positive definite matrix satisfying $s^2 = x'x$. [One version of this well known result is Proposition 5.5 in Eaton (1983).] The uniqueness of s in this factorization shows that $\mathbf{Y}_1$ intersects each orbit in exactly one point. The measurability is easily checked since $\mathbf{Y}_1$ is a relatively open subset of $\mathbf{X}$ and the map

$$t(x) = \begin{pmatrix} (x'x)^{1/2} \\ 0 \end{pmatrix} \in \mathbf{Y}_1$$

is continuous.

Now, consider $X \in \mathbf{X}$ which satisfies $\mathscr{L}(X) = \mathscr{L}(gX)$. Random matrices with this property are sometimes said to have left-orthogonally invariant distributions. Examples of such distributions on $\mathbf{X}$ are provided by taking X to have density (with respect to Lebesgue measure) on $\mathbf{X}$ of the form $q(x'x)$ where q is a nonnegative function defined on $p \times p$ positive definite matrices. For example

$$q(x'x) = \frac{|\Sigma|^{-n/2}}{(\sqrt{2\pi})^{np}} \exp\left[-\frac{1}{2} \operatorname{tr} x'x\Sigma^{-1} \right],$$

where Σ is $p \times p$ and positive definite, correspond to X with iid rows which are $N_p(0, \Sigma)$.

When $\mathscr{L}(X) = \mathscr{L}(gX)$, $g \in O_n$, Theorem 4.4 implies that

$$\mathscr{L}(X) = \mathscr{L}\left(U\begin{pmatrix} S \\ 0 \end{pmatrix} \right),$$

where U is uniform on O_n and is independent of S. The distribution of S is arbitrary over $p \times p$ positive definites.

The representation

$$x = g\begin{pmatrix} s \\ 0 \end{pmatrix}$$

can also be written

$$x = g\begin{pmatrix} I_p \\ 0 \end{pmatrix} s = \psi s,$$

where

$$\psi = g\begin{pmatrix} I_p \\ 0 \end{pmatrix}$$

is an element of $F_{p,n}$ as described in Example 2.3. Thus

$$\mathscr{L}(X) = \mathscr{L}\left(U\begin{pmatrix} S \\ 0 \end{pmatrix}\right) = \mathscr{L}\left(U\begin{pmatrix} I_p \\ 0 \end{pmatrix} S\right) = \mathscr{L}(\Delta S)$$

with Δ and S independent. That Δ has a uniform distribution on $F_{p,n}$ follows from the transitivity of the action of O_n on $F_{p,n}$.

Much the same analysis as above can be given based on the representation

$$x = g\begin{pmatrix} u \\ 0 \end{pmatrix},$$

where u is an element of the group of $p \times p$ upper triangular matrices with positive diagonal elements, say G_U^+. In this case, a cross section is taken to be

$$\mathbf{Y}_2 = \left\{\begin{pmatrix} u \\ 0 \end{pmatrix} \in \mathbf{X} \middle| u \in G_U^+\right\}.$$

That u is unique in this representation is well known [for example, see Proposition 5.2 in Eaton (1983)]. The remainder of the analysis and application of Theorem 4.4 is left to the reader. □

EXAMPLE 4.5. Let S_p be the real vector space of the $p \times p$ symmetric matrices. The group O_p acts on S_p via

$$s \to gsg'$$

for $s \in S_p$ and $g \in O_p$. To say that a random element $X \in S_p$ has an invariant distribution is to say that

$$\mathscr{L}(X) = \mathscr{L}(gXg'), \qquad g \in O_p.$$

Examples of such distributions on S_p include the Wishart distribution with identity scale matrix as well as certain versions of the multivariate beta and multivariate F distributions. For example, see Olkin and Rubin (1964). Define $\mathbf{Y} \subset S_p$ by

$$\mathbf{Y} = \{D \in S_p | D \text{ is diagonal, } d_{11} \geq \cdots \geq d_{pp}\},$$

where the diagonal elements of D are $d_{11}, \dots, d_{pp}$. Clearly $\mathbf{Y}$ is a closed subset of S_p. The spectral theorem shows that every $s \in S_p$ can be written

$$s = gDg'$$

for some $g \in O_p$ and some $D \in \mathbf{Y}$. Of course, the diagonal elements of D are the eigenvalues of s and the function

$$t(s) = D$$

is continuous and hence measurable. That **Y** is a measurable cross section is now apparent.

Theorem 4.4 asserts that if

$$\mathscr{L}(X) = \mathscr{L}(gXg') \quad \text{for } g \in O_p, \text{ then} \quad \mathscr{L}(X) = \mathscr{L}(UYU'),$$

where U is uniform on O_p and is independent of $Y \in \mathbf{Y}$. The distribution of Y is arbitrary. In other words X is generated by choosing the ordered eigenvalues according to some arbitrary distribution and then randomly "moving" Y via the map

$$Y \to UYU'$$

with U uniform on O_p. □

EXAMPLE 4.6. This example deals with yet another matrix decomposition result involving singular values. The space **X** is the vector space $\mathscr{L}_{p,n}$ of $n \times p$ real matrices with $p \le n$. The group in question is the product group $O_n \times O_p$ which acts on $\mathscr{L}_{p,n}$ via

$$x \to gxh'$$

for $g \in O_n$ and $h \in O_p$. The singular value decomposition for x is

$$x = g\begin{pmatrix} D \\ 0 \end{pmatrix} h$$

with $g \in O_n$ and $h \in O_p$. Here D is a $p \times p$ diagonal matrix with diagonal elements $d_{11} \ge \cdots \ge d_{pp} \ge 0$. These diagonal elements are the square roots of the eigenvalues of $x'x$. Thus, a candidate for a cross section is

$$\mathbf{Y} = \left\{ \begin{pmatrix} D \\ 0 \end{pmatrix} \in \mathscr{L}_{p,n} \middle| D \text{ is diagonal, } d_{11} \ge \cdots \ge d_{pp} \ge 0 \right\}.$$

Arguments similar to those given previously show that indeed **Y** is a measurable cross section.

Now, consider $X \in \mathscr{L}_{p,n}$ such that $\mathscr{L}(X) = \mathscr{L}(gXh')$ for $g \in O_n$ and $h \in O_p$. Examples of such distributions include the multivariate normal distribution on $\mathscr{L}_{p,n}$ (with mean 0 and identity covariance) and certain versions of the multivariate t distribution [for example, see Dickey (1967) or Eaton (1985)]. To describe the implications of Theorem 4.4, first note that the uniform distribution on $O_n \times O_p$ is just product Haar measure because $O_n \times O_p$ is a direct product. Thus, $U = (U_1, U_2)$ is uniform on $O_n \times O_p$ when U_1 is uniform on O_n, U_2 is uniform on O_p and U_1 and U_2 are independent. Therefore, when X has an invariant distribution on $\mathscr{L}_{p,n}$,

$$\mathscr{L}(X) = \mathscr{L}(U_1 Y U_2'),$$

where U_1, Y and U_2 are mutually independent and Y has an arbitrary distribution on **Y**. □

4.3. Null robustness applications. The material in this section comes mainly from Das Gupta (1979) and from Eaton and Kariya (1984). The problem discussed here is motivated by the following example. Consider a random vector

$X \in R^n$ and set

$$T(X) = \frac{e'X}{\|X\|},$$

where e is the vector of 1's in R^n and $\|\cdot\|$ denotes the usual norm on R^n. Student's t statistic is a one-to-one function of $T(X)$ so the distribution of $T(X)$ determines the distribution of Student's t statistic and conversely. In fact, a bit of algebra shows that

$$t_{n-1} = \frac{(n-1)^{1/2}T}{n^{1/2}(1 - n^{-1}T^2)^{1/2}},$$

where t_{n-1} denotes the usual t statistic. When the coordinates of X are iid $N(0,1)$, then t_{n-1} has the Student t_{n-1} distribution and hence the distribution of T is fixed, say Q_0. Now, we ask: Under what conditions on $\mathscr{L}(X)$ does $\mathscr{L}(T)$ remain fixed at Q_0 (and hence t_{n-1} will still have Student's t_{n-1} distribution)? Fisher observed that if $\mathscr{L}(X)$ is O_n-invariant, as it is when the coordinates of X are iid $N(0,1)$, then $X/\|X\|$ is uniform on $\{x | x \in R^n, \ \|x\| = 1\}$. Thus, the distribution of $T(X)$ must remain the same when the coordinates of X are iid $N(0,1)$ as when $\mathscr{L}(X)$ is O_n-invariant. What makes this argument tick is:

(i) $T(X) = T(cX)$, $c > 0$.
(ii) $X/\|X\|$ is uniform when $\mathscr{L}(X)$ is O_n-invariant.

Condition (i) implies T is a function of $X/\|X\|$ while (ii) fixes the distribution of $X/\|X\|$.

There are a number of other examples where arguments similar to the one above can be used to show that distributions of statistics of interest can be derived using invariance rather than distributional assumptions. In such cases, the conditions under which the statistic has the given distribution can sometimes be substantially weakened. It is this general problem to which we now turn.

Here is one way to describe the problem. A random variable $X \in \mathbf{X}$ is given as is a statistic $T(X)$. A compact group K acts measurably on $\mathbf{X}$. Let $\mathscr{P}_K$ denote the set of all probability measures on $\mathbf{X}$ which are K invariant. In what follows, $\mathscr{L}(T(X)|P)$ denotes the distribution of $T(X)$ when $\mathscr{L}(X) = P$. The problem addressed below is the following:

Under what conditions is it the case that

$$\mathscr{L}(T(X)|P) = \mathscr{L}(T(X)|P') \quad \text{for all } P, P' \in \mathscr{P}_K? \tag{4.10}$$

Under some regularity, Das Gupta (1979) provided some sufficient conditions for (4.10) to hold. To describe these, first assume that

$$T(x_1) = T(x_2) \quad \text{implies} \quad T(kx_1) = T(kx_2) \quad \text{for all } k \in K. \tag{4.11}$$

Then, by Theorem 2.3, the group action of K can be moved to the range space of T, say $(\mathbf{Y}, \mathscr{B}_1)$. Assume that K acts measurably on $(\mathbf{Y}, \mathscr{B}_1)$.

THEOREM 4.5. *If K acts transitively on $\mathbf{Y}$, then* (4.10) *holds.*

PROOF. The transitivity of K on $(\mathbf{Y}, \mathscr{B}_1)$ implies there is exactly one K invariant probability measure on $(\mathbf{Y}, \mathscr{B}_1)$, say Q_0. But, when $\mathscr{L}(X) \in \mathscr{P}_K$, then $\mathscr{L}(X) = \mathscr{L}(kX)$ for all $k \in K$. Using the definition of the induced group action, we see that $\mathscr{L}(X) \in \mathscr{P}_K$ implies

$$\mathscr{L}(T(X)) = \mathscr{L}(T(kX)) = \mathscr{L}(kT(X)).$$

Thus, the induced distribution of T is K invariant so $\mathscr{L}(T(X)) = Q_0$. □

EXAMPLE 4.7. Suppose $X_1, \ldots, X_n$ is a random sample from a p-dimensional $N(\mu, \Sigma)$ with μ and Σ unknown. The usual Hotelling statistic for testing $\mu = 0$ is easily shown to be a function of

$$T(X) = X(X'X)^{-1}X',$$

where X: $n \times p$ has rows $X_1', \ldots, X_n'$. When $\mu = 0$,

$$\mathscr{L}(X) = N(0, I_n \otimes \Sigma),$$

so $\mathscr{L}(\gamma X) = \mathscr{L}(X)$ for $\gamma \in O_n$. The sample space $\mathbf{X}$ for this example is taken to be the set of $n \times p$ real matrices of rank p (a set of Lebesgue measure 0 has been removed). Thus, the range of T is the set $S_{n,p}$ of $n \times n$ rank p orthogonal projections.

With $K = O_n$, assumption (4.11) is valid as is the transitivity of the induced group action on $S_{n,p}$. The conclusion is that $\mathscr{L}(T)$ is equal to the unique invariant probability on $S_{n,p}$ which was introduced in Example 2.10. Hence the null distribution of Hotelling's statistic is the same when $\mathscr{L}(X) = N(0, I_n \otimes \Sigma)$ as when $\mathscr{L}(X) = \mathscr{L}(\gamma X)$ for $\gamma \in O_n$. □

In some cases of interest, assumption (4.11) does not hold but (4.10) is still valid. An alternative set of assumptions which yields (4.10) is given in Eaton and Kariya (1984). To describe these, assume that G is a topological group which acts measurably and transitively on $\mathbf{X}$, K is a compact subgroup of G and H is a subgroup of G such that

$$G = K \cdot H = \{kh | k \in K, h \in H\}.$$

Therefore, the subgroups K and H generate G.

THEOREM 4.6. *Assume that $T(X)$ is an H-invariant function. If either H or K is a normal subgroup of G, then* (4.10) *holds.*

PROOF. First assume that K is normal in G. To establish the theorem, it suffices to show that for any bounded measurable function f,

$$\int f(T(x))P(dx) = \int f(T(x))P'(dx) \quad \text{for } P, P' \in \mathscr{P}_K. \tag{4.12}$$

Because $kP = P$, we have

$$\int f(T(x))P(dx) = \int f(T(kx))P(dx).$$

Integrating both sides of this equality with respect to the invariant probability measure on K yields

$$\int f(T(x))P(dx) = \iint f(T(kx))\nu(dk)P(dx).$$

Fix $x_0 \in \mathbf{X}$ and use the transitivity of $G = K \cdot H$ to write $x = k_1 h x_0$. Using the invariance of T under H and the invariance of ν, we have

$$\int f(T(kx))\nu(dx) = \int f(T(kk_1hx_0))\nu(dk) = \int f\big(T\big(h^{-1}khx_0\big)\big)\nu(dk).$$

Since the map $k \to h^{-1}kh$ is a continuous isomorphism of K, the uniqueness of ν implies that ν is invariant under this map. Therefore

$$\int f(T(kx))\nu(dk) = \int f(T(kx_0))\nu(dk),$$

which yields the equation

$$\int f(T(x))P(dx) = \int f(T(kx_0))\nu(dk).$$

Because this equation holds for each $P \in \mathscr{P}_K$, obviously (4.12) holds. Thus the theorem is proved when K is normal in G. The proof when H is normal in G is similar and the details are left to the reader. □

Example 4.8. In this example where canonical correlations are discussed, Theorem 4.6 is applicable but Theorem 4.5 is not. Consider a random matrix Z: $n \times p$ which has rank p and partition Z as $Z = (Z_1 Z_2)$, where Z_i is $n \times p_i$, $i = 1, 2$. Without essential loss of generality, the mean 0 case is treated here. The random rank p_i orthogonal projection

$$Q_i = Z_i(Z_i'Z_i)^{-1}Z_i', \qquad i = 1, 2,$$

takes values in the space $S_{n,\,p_i}$ of the last example. The *squared canonical correlations* are defined to be the $r = \min\{p_1, p_2\}$ largest eigenvalues of Q_1Q_2. That this definition agrees with more traditional definitions is easily checked.

Given Z, let $T(Z)$ be the vector of the r largest eigenvalues (arranged in order) of Q_1Q_2. When Z is $N(0, I_n \otimes I_p)$, the density of $T(Z)$ is given in Anderson [(1958), Chapter 13]. To describe a large class of distributions of Z for which the distribution of $T(Z)$ is that when Z is $N(0, I_n \otimes I_p)$, consider the group G whose elements are (γ, ψ, A, B) with $\gamma, \psi \in O_n$, $A \in \mathrm{Gl}_{p_1}$ and $B \in \mathrm{Gl}_{p_2}$. The action of G on (z_1, z_2) is

$$(z_1, z_2) \to (\gamma z_1 A', \psi z_2 B')$$

and the group operation is

$$(\gamma_1, \psi_1, A_1, B_1)(\gamma_2, \psi_2, A_2, B_2) = (\gamma_1\gamma_2, \psi_1\psi_2, A_1A_2, B_1B_2).$$

That G is transitive on the set $\mathbf{X}$ of $n \times p$ matrices of rank p is easily checked.

To apply Theorem 4.6, let

$$H = \{\gamma, \psi, A, B)|\gamma = \psi\}$$

and

$$K = \{(\gamma, \psi, A, B) | \psi = I_n, A = I_{p_1}, B = I_{p_2}\}.$$

Then $G = K \cdot H$, K is compact and K is normal in G. Since $T(Z)$ is H-invariant, (4.10) holds. In other words,

$$\mathscr{L}((Z_1, Z_2)) = \mathscr{L}((\gamma Z_1, Z_2))$$

implies that the distribution of $T(Z)$ is the same as when $\mathscr{L}(Z) = N(0, I_n \otimes I_p)$. □

Other examples and references can be found in Das Gupta (1979) and Eaton and Kariya (1984).

CHAPTER 5

Decomposable Measures

The main purpose of this chapter is to discuss the extent to which Theorem 4.3 can be generalized to arbitrary Radon measures (rather than probability measures) and to cases where G is not compact. Such generalizations can be used in the derivation of densities of maximal invariants as well as in other areas. The approach here is modelled after that described in Andersson (1982). Other possible approaches to this problem are described in Wijsman (1986) (the global cross section approach using some Lie group theory) and Farrell (1985) (a measure-theoretic cross section approach developed by Schwartz (1966), unpublished).

The general method of averaging over a group to obtain a density of a maximal invariant is due to Stein (1956). However, there are mathematical problems to overcome. The approaches described in Farrell (1985) and Wijsman (1986) have their advantages and disadvantages as does the method to be described here. As far as I know, there are no "practical" problems where one of the methods can be applied but the others cannot. The ease with which the methods apply depends on the problem at hand and the method most familiar to you. It should be noted however that all of the methods require some regularity conditions which are essential.

In the first section of this chapter, we treat the compact group case. A version of Theorem 4.3 is established for Radon measures. However the methods and the language are quite different here because the methods used in Chapter 4 do not carry over easily to the case when the group is noncompact. The noncompact case is discussed in Section 5.2. In Section 5.3, a representation of the density of a maximal invariant due to Andersson (1982) is established. This representation is used to provide a proof of an important result due to Wijsman (1967) on ratios of densities of a maximal invariant statistic.

5.1. The compact group case. Throughout this section $\mathbf{X}$ is a locally compact Hausdorff space with a countable base for the topology (so the topology is a metric topology). Also, G is a compact topological group which acts topologically on $\mathbf{X}$. Thus, the map $(g, x) \to gx$ from $G \times \mathbf{X}$ to $\mathbf{X}$ is continuous.

Rather than introduce a particular measurable cross section, we will consider the quotient space $\mathbf{X}/G$ whose points are the equivalence classes $\{gx|g \in G\}$. That is, the points in $\mathbf{X}/G$ are just the orbits. The natural projection π on $\mathbf{X}$ to $\mathbf{X}/G$ given by

$$\pi(x) = G \cdot x = \{gx|g \in G\}$$

plays an important role in what follows. Observe that π is a maximal invariant function. Thus, a real valued function f defined on $\mathbf{X}$ is invariant iff there exists a real valued function f^* on $\mathbf{X}/G$ such that

$$f(x) = f^*(\pi(x)). \tag{5.1}$$

Let ν denote the unique invariant probability measure on G. Temporarily ignoring a host of technical considerations, look at the function

$$x \to \int_G f(gx)\nu(dg).$$

This function is clearly G-invariant, so it can be thought of as a map which sends f into a function f^* defined on $\mathbf{X}/G$. Thus, Theorem 4.3 can be written

$$\int_{\mathbf{X}} f(x)P(dx) = \int_{\mathbf{X}/G} f^*(y)Q(dy).$$

Equivalently, if J is the invariant integral defined by an invariant P and J_1 is the integral defined by Q, the above is

$$J(f) = J_1(T(f)), \tag{5.2}$$

where $T(f) = f^*$ is the mapping which sends f into f^*. Theorem 4.3 shows that if J is an invariant integral (corresponding to a probability measure), then there exists an integral J_1 (corresponding to a probability measure on $\mathbf{X}/G$) such that (5.2) holds. The validity of (5.2) for an arbitrary invariant integral J is the question to which we now turn.

Here is the general plan of attack. First, when $\mathbf{X}/G$ has the quotient topology (which makes $\mathbf{X}/G$ a locally compact Hausdorff space), it will be shown that the mapping T described above maps $K(\mathbf{X})$ *onto* $K(\mathbf{X}/G)$. Thus, given an invariant integral J defined on $K(\mathbf{X})$, we *define* J_1 on $K(\mathbf{X}/G)$ via the equation

$$J(f) = J_1(T(f)), \qquad f \in K(\mathbf{X}).$$

Of course, there is some work involved in showing that J_1 is well defined and J_1 is an integral. The uniqueness of J_1 follows because T is onto and hence (5.2) holds. We now turn to the technical details.

First, a few words about the quotient topology for $\mathbf{X}/G$. Recall that $\mathbf{X}$ is a locally compact Hausdorff space with a countable base for the topology and G is a compact topological group with a countable base for its topology. The action $(g, x) \to gx$ is assumed to be continuous on $G \times \mathbf{X}$ to $\mathbf{X}$. A subset $U \subset \mathbf{X}/G$ is open (in the quotient topology) iff $\pi^{-1}(U)$ is open in $\mathbf{X}$ where π is the natural projection on $\mathbf{X}$ to $\mathbf{X}/G$. Because G is compact and the topology for $\mathbf{X}$ has a countable base, the quotient topology (i) is a locally compact Hausdorff topology and (ii) has a countable base. Thus, the quotient space $\mathbf{X}/G$ is of the same type as the space $\mathbf{X}$.

Next, we consider the function T defined on $K(\mathbf{X})$ by

$T(f)$ is the unique function f^* defined on $\mathbf{X}/G$ which satisfies

$$\int_G f(gx)\nu(dg) = f^*(\pi(x)).$$

THEOREM 5.1. *The function T maps $K(\mathbf{X})$ onto $K(\mathbf{X}/G)$. Also, T satisfies*

$$T(\alpha f_1 + \beta f_2) = \alpha T(f_1) + \beta T(f_2) \tag{5.3}$$
$$T(f) \geq 0 \quad \text{when } f \geq 0,$$

for $\alpha, \beta \in R^1$ and $f \in K(\mathbf{X})$.

PROOF. For $f \in K(\mathbf{X})$, the continuity of the function

$$x \to \int_G f(gx)\nu(dg)$$

is easily established using the bounded convergence theorem. Thus, by definition of the quotient topology, f^* is continuous. To show f^* has compact support when f has compact support $V \subset \mathbf{X}$, first note that the set

$$G \cdot V = \{y | y = gx \text{ for some } g \in G,\ x \in V\}$$

is a compact subset of $\mathbf{X}$ because $G \cdot V$ is the continuous image of the compact set $G \times V \subset G \times \mathbf{X}$. Thus $\pi(G \cdot V)$ is a compact set in $\mathbf{X}/G$ since π is continuous. If $y \notin \pi(G \cdot V)$, then $\pi^{-1}(y) = G \cdot x$ for some $x \notin V$. Thus for all $g \in G$, $f(gx) = 0$ because $gx \notin G \cdot V$ and f vanishes outside $G \cdot V$. Therefore $f^*(y) = 0$ for y outside the compact set $\pi(G \cdot V)$.

To show T is onto, let $f^* \in K(\mathbf{X}/G)$ and consider

$$f(x) = f^*(\pi(x)), \qquad x \in \mathbf{X}.$$

That f is continuous is obvious. If the compact set $V \subset \mathbf{X}/G$ contains the support of f^*, then $\pi^{-1}(V)$ is easily shown to be compact in $\mathbf{X}$. Obviously, if $x \notin \pi^{-1}(V)$, then $f(x) = 0$ so $\pi^{-1}(V)$ supports f. Because f defined above is G-invariant,

$$\int f(gx)\nu(dg) = f(x) = f^*(\pi(x)),$$

so $T(f) = f^*$. Hence T is onto.

That T satisfies the relations in (5.3) is easily proved. This completes the proof. □

The results of Theorem 5.1 provide the key technical step in the following generalization of Theorem 4.3.

THEOREM 5.2. *Suppose J is a G-invariant integral on $K(\mathbf{X})$. Then there exists a unique integral J_1 on $K(\mathbf{X}/G)$ such that*

$$J(f) = J_1(T(f)), \qquad f \in K(\mathbf{X}). \tag{5.4}$$

Conversely, for each integral J_1 on $K(\mathbf{X}/G)$, the integral J on $K(\mathbf{X})$ defined by

$$J(f) = J_1(T(f))$$

is G-invariant.

PROOF. Given $f^* \in K(\mathbf{X}/G)$, define J_1 by

$$J_1(f^*) = J(f)$$

for any f such that $T(f) = f^*$. To show that J_1 is well defined, suppose $T(f_1) = T(f_2) = f^*$. With $f_3 = f_1 - f_2$, obviously $T(f_3) = 0$. It must be shown that this implies $J(f_3) = 0$. First, represent J by its associated Radon measure μ:

$$J(f) = \int_{\mathbf{X}} f(x)\mu(dx).$$

The invariance of J yields

$$J(f) = \int_{\mathbf{X}} f(gx)\mu(dx), \qquad g \in G,$$

so that

$$J(f) = \int_{\mathbf{X}} \int_G f(gx)\nu(dg)\mu(dx).$$

Thus, if $T(f_3) = 0$,

$$T(f_3)(\pi(x)) = \int_G f_3(gx)\nu(dg) = 0,$$

which shows that $J(f_3) = 0$. Hence J_1 is well defined. The linearity of J_1 and the inequality $J_1(f^*) \geq 0$ for $f^* \geq 0$ are easily established. Hence J_1 is an integral and (5.4) holds. The uniqueness of J_1 follows from the fact that T is an onto map.

For the converse, just observe that $T(f) = T(L_g f)$ for $f \in K(\mathbf{X})$ and $g \in G$. Hence J defined by

$$J(f) = J_1(T(f))$$

is G-invariant. □

When the relationship (5.4) holds for $f \in K(\mathbf{X})$, then of course (5.4) holds for all integrable functions f because both sides of (5.4) are integrals. Thus, under the assumptions of Theorem 5.2, Equation (5.4) can be used for integrable functions as well as functions in $K(\mathbf{X})$.

Our first application of Theorem 5.2 involves finding the density function of a maximal invariant. Consider X taking values in a space $\mathbf{X}$ and assume that X has a density p with respect to a Radon measure μ.

THEOREM 5.3. *Assume that the dominating measure μ is invariant under the compact group G and let J denote the integral defined by μ. Write $J = J_1 \circ T$*

as in Equation (5.4) *and let* μ_1 *be the Radon measure associated with* J_1 *defined on the quotient space. Then the density function of* $Y = \pi(X)$ *with respect to* μ_1 *is* $p^* = T(p)$.

PROOF. Let f^* be any bounded measurable function defined on $\mathbf{X}/G$. It suffices to show that

$$\mathscr{E}f^*(Y) = \int_{\mathbf{X}/G} f^*(y)p^*(y)\mu_1(dy).$$

First observe that $f^*\pi$ is a bounded measurable function defined on $\mathbf{X}$ and $T((f^*\pi)p) = f^*T(p)$. Therefore,

$$\begin{aligned}\mathscr{E}f^*(Y) &= \mathscr{E}f^*(\pi(X)) = J((f^*\pi)p) = J_1(T((f^*\pi)p)) \\ &= J_1(f^*T(p)) = J_1(f^*p^*) = \int f^*(y)p^*(y)\mu_1(dy),\end{aligned}$$

where the third equality follows from Equation (5.4) in Theorem 5.2. This completes the proof. □

The application of Theorem 5.3 requires two separate steps—the calculation of the induced measure μ_1 and the evaluation of p^* which involves the calculation of

$$\int_G p(gx)\nu(dg).$$

In concrete problems, one often chooses a particular representation of the quotient space $\mathbf{X}/G$ [that is, some one-to-one function of $\pi(x)$] to facilitate the discussion of the density of a maximal invariant. In symbols, suppose $k: \mathbf{X}/G \to Z$ is a one-to-one onto bimeasurable function. Thus $Z = k(\pi(X))$ is a maximal invariant. Let δ be the image of μ_1 under k, that is, $\delta(B) = \mu_1(k^{-1}(B))$. A direct calculation shows that $\tilde{p}(z) = p^*(k^{-1}(z))$ is the density of Z with respect to δ.

The implication of this is that the calculation of the density of *any* maximal invariant function, say $k_0(X) = W$, involves (i) the calculation of the induced measure μ_0 given by $\mu_0(B) = \mu(k_0^{-1}(B))$ and (ii) writing the function

$$x \to \int p(gx)\nu(dg)$$

as a function of $w = k_0(x)$, say $p_0(w)$.

Here is the classical example of the noncentral Wishart density [Herz (1955) and James (1954)] to which the above argument applies.

EXAMPLE 5.1. Let $\mathbf{X}$ be the space of $n \times p$ real matrices of rank p and take μ to be Lebesgue measure on $\mathbf{X}$. The group $G = O_n$ acts on $\mathbf{X}$ via matrix multiplication:

$$x \to gx, \qquad g \in O_n.$$

Obviously μ is an invariant measure. A standard choice for a maximal invariant is

$$s = k_0(x) = x'x,$$

so s takes values in S_p^+—the set of $p \times p$ positive definite real matrices. To find the induced measure μ_0 on S_p^+ defined by

$$\mu_0(B) = \mu\big(k_0^{-1}(B)\big),$$

first observe that μ_0 is characterized by the equation

$$\int f^*(k_0(x))\,dx = \int f^*(s)\mu_0(ds), \tag{5.5}$$

which holds for all bounded integrable f^* on S_p^+. To solve (5.5) for μ_0, take f^* to be of the form

$$f^*(k_0(x)) = f_1^*(k_0(x))\phi(k_0(x)), \tag{5.6}$$

where

$$\phi(s) = (\sqrt{2\pi})^{-np}\exp\left[-\tfrac{1}{2}\operatorname{tr} s\right].$$

Then $\phi(k_0(x))$ is the density of a normal distribution on $\mathbf{X}$ (with mean zero and covariance equal to the identity). Thus, the left side of (5.5) is

$$\int f_1^*(k_0(X))\phi(k_0(x))\,dx$$

which is the expectation of

$$f_1^*(k_0(X)) = f_1^*(S),$$

where $S = X'X$. Now, standard multivariate arguments show that the density of S (with respect to Lebesgue measure ds on S_p^+) is

$$p_1^*(s) = \omega(n,p)|s|^{(n-p-1)/2}\exp\left[-\tfrac{1}{2}\operatorname{tr} s\right],$$

where $\omega(n, p)$ is the Wishart constant

$$[\omega(n,p)]^{-1} = \pi^{p(p-1)/4}2^{np/2}\prod_{j=1}^{p}\Gamma\left(\frac{n-j+1}{2}\right).$$

Thus, the right side of (5.5) is just the expectation of $f_1^*(S)$ relative to the density $p_1^*(s)$. This yields the equation

$$\int f_1^*(s)\phi(s)\mu_0(ds) = \int f_1^*(s)p_1^*(s)\,ds,$$

which holds for all bounded measurable f_1^*. Therefore

$$\mu_0(ds) = \frac{p_1^*(s)}{\phi(s)}\,ds = \omega(n,p)(\sqrt{2\pi})^{np}|s|^{(n-p-1)/2}\,ds. \tag{5.7}$$

Now, let $p(x)$ be a density of X with respect to Lebesgue measure on $\mathbf{X}$. Theorem 5.3 and the succeeding discussion show that the density of $X'X = S$ with respect to μ_0 is calculated by evaluating the integral

$$\int_{O_n} p(gx)\nu(dg)$$

and writing the answer as a function of $s \in S_p^+$. For this particular case, x can be written (uniquely) as

$$x = h\begin{pmatrix} s^{1/2} \\ 0 \end{pmatrix}$$

where $s = x'x$ is in S_p^+ and $h \in O_n$. Therefore the density of S is

$$p^*(s) = \int_{O_n} p\left(gh\begin{pmatrix} s^{1/2} \\ 0 \end{pmatrix}\right)\nu(dg) = \int_{O_n} p\left(g\begin{pmatrix} s^{1/2} \\ 0 \end{pmatrix}\right)\nu(dg). \tag{5.8}$$

The particular choice of p which leads to the noncentral Wishart distribution is

$$p_0(x) = (\sqrt{2\pi})^{-np} \exp\left[-\tfrac{1}{2}\operatorname{tr}(x-\theta)'(x-\theta)\right], \tag{5.9}$$

where θ is an $n \times p$ matrix. Of course this choice of p_0 corresponds to X having a $N(\theta, I_n \otimes I_p)$ distribution. Substitution of (5.9) into (5.8) yields

$$p^*(s) = (\sqrt{2\pi})^{-np} \exp\left[-\tfrac{1}{2}\operatorname{tr} s - \tfrac{1}{2}\operatorname{tr}\theta'\theta\right] \int_{O_n} \exp[\operatorname{tr} gz]\nu(dg), \tag{5.10}$$

where

$$z = \begin{pmatrix} s^{1/2} \\ 0 \end{pmatrix}\theta'.$$

Thus, the difficulty is the evaluation of

$$\psi(z) = \int_{O_n} \exp[\operatorname{tr} gz]\nu(dg). \tag{5.11}$$

It is the attempted evaluation of (5.11) which led to the development of zonal polynomials by Alan James and others. The reader can find an excellent discussion of zonal polynomials in Muirhead (1982), Farrell (1985) and Takemura (1984). This subject is not discussed further here. □

5.2. The noncompact case. In this section, we discuss the validity of Theorem 5.2 when the group G is not necessarily compact. Throughout this section, the group G is assumed to be a locally compact, σ-compact topological group whose topology has a countable base. The basic approach here is to make some modifications in the method of proof used in the compact case so that versions of both Theorems 5.1 and 5.2 become valid for noncompact groups.

The first problem to overcome concerns the appropriate definition of the function T discussed in Theorem 5.1. Thus, consider the group G acting topologically on the space $\mathbf{X}$. Let ν_l denote left Haar measure on G and let Δ be the modular function of G. Then

$$\nu_r(dg) = \frac{1}{\Delta(g)}\nu_l(dg)$$

is a right Haar measure on G. First observe that for $f \in K(\mathbf{X})$, the function

$$x \to \int f(gx)\nu_r(dg) \tag{5.12}$$

is an invariant function of x—assuming the integral makes sense. (We will discuss some conditions under which the integral is well defined a bit later, but for now assume everything is all right.) Thus, (5.12) provides the appropriate definition of the function T which supposedly maps $K(\mathbf{X})$ into $K(\mathbf{X}/G)$ (assuming the quotient space is a "reasonable" space and ignoring the continuity and compact support questions). Therefore $T(f)$ is defined to be the function f^* on $\mathbf{X}/G$ which satisfies

$$\int f(gx)\nu_r(dg) = f^*(\pi(x)), \tag{5.13}$$

where π is the natural projection from $\mathbf{X}$ to the quotient space $\mathbf{X}/G$ given by

$$\pi(x) = G \cdot x = \{gx | g \in G\}.$$

If J_1 is an integral defined on $K(\mathbf{X}/G)$, then the expression $J_1(T(f))$ should define an integral on $K(\mathbf{X})$. But, for $h \in G$,

$$T(L_h f) = \Delta^{-1}(h)T(f)$$

because

$$\begin{aligned}
\int (L_h f)(gx)\nu_r(dg) &= \int f(h^{-1}gx)\nu_r(dg) \\
&= \int f(h^{-1}gx)\Delta^{-1}(g)\nu_l(dg) \\
&= \Delta^{-1}(h)\int f(h^{-1}gx)\Delta^{-1}(h^{-1}g)\nu_l(dg) \\
&= \Delta^{-1}(h)\int f(gx)\Delta^{-1}(g)\nu_l(dg) \\
&= \Delta^{-1}(h)\int f(gx)\nu_r(dg).
\end{aligned}$$

Hence, if $J_1(T(f))$ does define an integral on $K(\mathbf{X})$, this integral is relatively invariant with multiplier Δ^{-1}. The whole point of this discussion is that the natural definition of T in the noncompact case leads to integrals $J_1(T(f))$ which are relatively invariant with multiplier Δ^{-1}. *Therefore, the only types of integrals J which can possibly have the representation $J(f) = J_1(T(f))$ must be relatively invariant with multiplier Δ^{-1}, given our definition of T.* This issue did not arise in the compact case because there are no nontrivial multipliers when the group is compact.

Now, consider G acting on $\mathbf{X}$ and suppose the integral J defined on $K(\mathbf{X})$ is relatively invariant with multiplier Δ^{-1}. We now want to discuss conditions under which a representation for J of the above type holds. Here is a simple example which shows that some additional assumptions need to be made.

EXAMPLE 5.2. Let $\mathbf{X} = R^1$ and take $G = R^2$ with addition as the group operation. For $g = (a_1, a_2) \in R^2$, the action of G on $\mathbf{X}$ is defined by

$$(a_1, a_2)x = x + a_1.$$

Obviously Lebesgue measure on $\mathbf{X}$ is invariant under this group action. Because G is a commutative group, the modulus of G is identically 1 and Lebesgue measure dg on G is right- and left-invariant. The difficulty here arises with the definition of T. Consider $f \geq 0$, $f \in K(\mathbf{X})$ and form

$$\int_G f(gx)\,dg = \int\int f(x + a_1)\,da_1\,da_2.$$

This integral is $+\infty$ as long as f is not 0, and there is just no way to patch things. The problem is that when $f \geq 0$ is not 0, for each x, the set

$$\{g | f(gx) \geq \varepsilon\}$$

has infinite measure for some $\varepsilon > 0$. This sort of situation must be ruled out in order to have a representation theorem. Because G acts transitively on $\mathbf{X}$ in this example, the trouble is not with the quotient space (or its topology), but the trouble is the size of the group compared to the size of the space. Indeed, for every interval $[c, d] \subset R^1$ with $c < d$ and for every $x \in R^1$,

$$\{g | gx \in [c, d]\} = \{(a_1, a_2) | a_1 \in [c - x, d - x]\} = [c - x, d - x] \times R^1$$

has infinite measure in G.

In order that the integral in (5.13)

$$\int_G f(gx)\nu_r(dg)$$

be well defined, it is sufficient that sets of the form

$$\{g | gx \in C\}$$

be compact when $C \subset \mathbf{X}$ is compact. This type of condition leads to a version of Theorem 5.2. □

Here is the condition which excludes the sort of situation encountered in Example 5.1. Consider the topological group G which acts topologically on the space $\mathbf{X}$.

DEFINITION 5.1. The group G acts *properly* on $\mathbf{X}$ if the mapping ψ defined on $G \times \mathbf{X}$ to $\mathbf{X} \times \mathbf{X}$ by

$$\psi(g, x) = (gx, x)$$

is a *proper* mapping, that is, if the inverse image under ψ of each compact set in $\mathbf{X} \times \mathbf{X}$ is a compact set in $G \times \mathbf{X}$.

Here is the analog of Theorem 5.1:

THEOREM 5.4. *Assume that G acts properly on* $\mathbf{X}$. *Then the quotient space* $\mathbf{X}/G$ *with its quotient topology is a locally compact Hausdorff space with a countable base for open sets. The mapping T on* $K(\mathbf{X})$ *given by* $T(f) = f^*$, *where* f^* *is the unique function satisfying*

$$\int f(gx)\nu_r(dg) = f^*(\pi(x)),$$

is well defined. Further T maps $K(\mathbf{X})$ *onto* $K(\mathbf{X}/G)$ *and satisfies condition* (5.3).

The above theorem is a conglomeration of results from Bourbaki. See Andersson (1982) for further discussion and references to the relevant portions of Bourbaki. Also, see the discussion in Wijsman (1985).

THEOREM 5.5. *Assume G acts properly on $\mathbf{X}$ and suppose that J is a relatively invariant integral on $K(\mathbf{X})$ with multiplier Δ^{-1}. Then there exists a unique integral J_1 on $K(\mathbf{X}/G)$ such that*

$$J(f) = J_1(T(f)), \qquad f \in K(\mathbf{X}). \tag{5.14}$$

Conversely, for each integral J_1 defined on $K(\mathbf{X}/G)$, J defined by (5.14) *is relatively invariant with multiplier Δ^{-1}.*

PROOF. The proof is essentially the same as the proof of Theorem 5.2 once it is noted that

$$T(L_h f) = \Delta^{-1}(h)T(f).$$

This relation was established earlier. The details are left to the reader. □

Before turning to some applications of Theorem 5.5, we first discuss a useful sufficient condition that G acts properly on $\mathbf{X}$. This condition also has the advantage of being a bit easier to understand than the condition of properness.

Given two subsets $A, B \subset \mathbf{X}$, set

$$(A, B) = \{g \in G | gA \cap B \neq \phi\}.$$

THEOREM 5.6. *Assume that for any compact subsets $A, B \subset \mathbf{X}$, the set (A, B) is a compact subset of G. Then G acts properly on $\mathbf{X}$.*

This theorem, which comes from Bourbaki, is given as Lemma 1.1 in Wijsman (1985) where a proof can be found. The assumption that (A, B) is compact whenever A and B are compact has some interesting consequences. For example, take $A = B = \{x\}$. Then

$$(\{x\}, \{x\}) = \{g | gx = x\},$$

which is the isotropy subgroup of x. Hence all of the isotropy subgroups are compact under the assumption of Theorem 5.6. When G is a "nice" subset of some Euclidean space, then the assumption that (A, B) is compact just means that (A, B) is a bounded set in the Euclidean space. (A, B) is always closed because the group action is continuous. Thus, in this case, the compactness of (A, B) is equivalent to the assertion that the collection of g's which "move" some point in A into B is bounded.

Finally, we turn to the question of a representation theorem for integrals (Radon measures) which are relatively invariant with a multiplier χ_0 on G. Thus suppose an integral J_0 on $K(\mathbf{X})$ is given by

$$J_0(f) = \int f(x)\mu_0(dx)$$

and

$$J_0(L_h f) = \int f(h^{-1}x)\mu_0(dx) = \chi_0(h)\int f(x)\mu_0(dx).$$

When G acts properly on $\mathbf{X}$ and when $\chi_0 = \Delta^{-1}$, then and only then does Theorem 5.5 apply directly. However, when $\chi_0 \neq \Delta^{-1}$, it is possible to change the measure μ_0 into a new measure to which Theorem 5.5 applies. Here are the details of that modification. Because the group G is assumed to be σ-compact, it follows that given any multiplier χ on G there exists a positive continuous function ϕ defined on $\mathbf{X}$ which satisfies

$$\phi(gx) = \chi(g)\phi(x) \tag{5.15}$$

for $x \in \mathbf{X}$ and $g \in G$. [See Bourbaki (1963), Proposition 7, Section 2, 4°.] Now, define a new measure μ by

$$\mu(dx) = \frac{1}{\phi(x)}\mu_0(dx), \tag{5.16}$$

where ϕ satisfies (5.15) with $\chi = \chi_0\Delta$. The claim is that the integral J defined by

$$J(f) = \int f(x)\mu(dx)$$

is relatively invariant with multiplier Δ^{-1}. This claim is verified as follows. For $h \in G$,

$$\begin{aligned}
J(L_h f) &= \int f(h^{-1}x)\mu(dx) = \int f(h^{-1}x)\frac{1}{\phi(x)}\mu_0(dx) \\
&= \int f(h^{-1}x)\frac{1}{\phi(hh^{-1}x)}\mu_0(dx) \\
&= \frac{1}{\chi_0(h)\Delta(h)}\int f(h^{-1}x)\frac{1}{\phi(h^{-1}x)}\mu_0(dx) \\
&= \frac{\chi_0(h)}{\chi_0(h)\Delta(h)}\int f(x)\frac{1}{\phi(x)}\mu_0(dx) = \Delta^{-1}(h)J(f).
\end{aligned}$$

Thus, under the assumption that G acts properly on $\mathbf{X}$, Theorem 5.5 can be applied to J. Summarizing this discussion gives:

THEOREM 5.7. *Assume G acts properly on $\mathbf{X}$ and let*

$$J_0(f) = \int f(x)\mu_0(dx)$$

be relatively invariant with multiplier χ_0. Pick ϕ to be a positive continuous function on $\mathbf{X}$ satisfying (5.15) *with $\chi = \chi_0\Delta$. Then the integral*

$$J(f) = \int f(x)\frac{1}{\phi(x)}\mu_0(dx)$$

is relatively invariant with multiplier Δ^{-1}. *Further* J_0 *has the representation*

$$J_0(f) = J(f\phi) = J_1(T(f\phi)) \tag{5.17}$$

for some integral J_1 *defined on* $K(\mathbf{X}/G)$.

PROOF. This follows immediately from the discussion above and Theorem 5.5. □

It should be mentioned that $T(f\phi)$ in (5.17) takes a special form because ϕ satisfies (5.15). Recall that for $f_1 \in K(\mathbf{X})$, $T(f_1) = f_1^*$ is the function in $K(\mathbf{X}/G)$ which satisfies the equation

$$\int f_1(gx)\nu_r(dg) = f_1^*(\pi(x)).$$

Substituting $f_1 = f\phi$ and using (5.15) yields

$$\begin{aligned}(f\phi)^*(\pi(x)) &= \int f(gx)\phi(gx)\nu_r(dg)\\ &= \phi(x)\int f(gx)\chi_0(g)\Delta(g)\nu_r(dg) \qquad (5.18)\\ &= \phi(x)\int f(gx)\chi_0(g)\nu_l(dg),\end{aligned}$$

where $\nu_l = \Delta\nu_r$ is a left Haar measure on G.

5.3. The Wijsman representation. The main result in this section is a version of the Wijsman representation for the ratio of densities of a maximal invariant. The version presented here, under the assumption that G acts properly on $\mathbf{X}$, is due to Andersson (1982) and is an easy consequence of Equation (5.18) following Theorem 5.7.

Our first goal is to establish a version of Theorem 5.3 for the noncompact case. Throughout this section, G is a locally compact, σ-compact topological group which acts topologically on $\mathbf{X}$. *It is also assumed that the action of G on $\mathbf{X}$ is proper, so Theorems* 5.5 *and* 5.6 *are valid.* As usual π denotes the natural projection of $\mathbf{X}$ onto $\mathbf{X}/G$.

THEOREM 5.8. *Consider a random variable X with values in $\mathbf{X}$ and assume that X has a density p with respect to a measure μ_0 which is relatively invariant with multiplier χ_0. As in Theorem* 5.7, *let ϕ be a positive continuous function satisfying* (5.15) *with $\chi = \chi_0\Delta$ so that the equation*

$$J_0(f) = J_1(T(f\phi)) \tag{5.19}$$

holds for some integral

$$J_1(f^*) = \int_{\mathbf{X}/G} f^*(y)\mu_1(dy)$$

defined on $K(\mathbf{X}/G)$. Then the density function of the maximal invariant $Y =$

$\pi(X)$ with respect to μ_1 is

$$(p\phi)^* = T(p\phi).$$

PROOF. The argument is a minor variation of that given in the proof of Theorem 5.3. It suffices to show that for $f^* \in K(\mathbf{X}/G)$,

$$\mathscr{E}f^*(Y) = \int f^*(y)(p\phi)^*(y)\mu_1(dy).$$

This equality follows from

$$\begin{aligned}\mathscr{E}f^*(Y) &= \mathscr{E}f^*(\pi(X))\\ &= J_0((f^*\pi)p) = J_1(T((f^*\pi)p\phi))\\ &= J_1(f^*T(p\phi)) = J_1(f^*(p\phi)^*)\\ &= \int f^*(y)(p\phi)^*(y)\mu_1(dy).\end{aligned}$$

This completes the proof. □

Here is Andersson's (1982) version of Wijsman's theorem:

THEOREM 5.9. *Under the assumptions of Theorem 5.8, let p_1 and p_2 be two possible densities of X. Then for each $x \in \mathbf{X}$ such that the denominator is positive, the ratio of the densities $(p_2\phi)^*/(p_1\phi)^*$ of the maximal invariant $\pi(X)$ is*

$$r(\pi(x)) = \frac{\int p_2(gx)\chi_0(g)\nu_l(dg)}{\int p_1(gx)\chi_0(g)\nu_l(dg)}. \tag{5.20}$$

PROOF. According to Theorem 5.8, when X has density p_i, the density of $\pi(X)$ with respect to μ_1 is $(p_i\phi)^*$. However, Equation (5.18) shows that

$$(p_i\phi)^*(\pi(x)) = \phi(x)\int p_i(gx)\chi_0(g)\nu_l(dg).$$

With division of $(p_2\phi)^*(\pi(x))$ by $(p_1\phi)^*(\pi(x))$, the result follows. □

Applications of this result to robustness and decision theory results are given in later chapters. The paper of Wijsman (1985) contains some very useful methods of verifying that certain group actions are proper. This paper also contains a number of important examples to which we refer later.

CHAPTER 6

Invariant Decision Problems

This rather lengthy chapter provides an introduction to invariant decision problems. After describing the basic ingredients in a decision problem, invariance is introduced and used to define an invariant decision problem. A main result in this chapter shows how to construct a best invariant rule when the group action is transitive on the parameter space and when the dominating measure is decomposable [That is, the integral J defined by the measure satisfies Equation (5.14) in Theorem 5.5.] Applications of this result to the construction of best invariant estimators are given.

Finally, invariant testing problems are discussed. Wijsman's theorem (described in Section 5.3) is used to derive an invariant test with some optimum properties.

6.1. Decision problems and invariance. In this section the basic objects of a decision problem are first reviewed and then invariance is introduced into the problem. Here are the ingredients of a decision problem:

(i) A sample space $(\mathbf{X}, \mathbf{B}_1)$.

(ii) A parameter space $(\Theta, \mathbf{B}_2)$.

(iii) An action space $(A, \mathbf{B}_3)$.

(iv) A statistical model $\{P_\theta | \theta \in \Theta\}$ which consists of a family of probability measures defined on the sample space.

(v) A loss function L defined on $A \times \Theta$ to $[0, \infty)$ and assumed to be jointly measurable.

To describe the decision rules, let $M_1(A)$ denote the set of all probability measures on the action space.

DEFINITION 6.1. A *decision rule* δ is a function defined on $\mathbf{X}$ with values in $M_1(A)$ such that δ is a Markov kernel (as in Example 2.19).

The value of a decision rule δ at x is denoted by $\delta(\cdot|x)$ and because δ is assumed to be a Markov kernel, the map

$$x \to \delta(B|x)$$

is Borel measurable for each fixed set $B \in \mathbf{B}_3$. In some of the literature δ is called a randomized decision rule since $\delta(\cdot|x)$ is a probability measure on the action space for each $x \in \mathbf{X}$ [for example, see Berger (1985)]. The decision rule δ is called *nonrandomized* if for each x, $\delta(\cdot|x)$ puts probability 1 at a single point, say $a(x)$, in A. It is easy to show that when δ is a nonrandomized rule, then the corresponding induced function $x \to a(x)$ is measurable on $(\mathbf{X}, \mathbf{B}_1)$ to $(A, \mathbf{B}_3)$. Conversely given a measurable function $x \to a(x)$, the corresponding δ which puts probability 1 at $a(x)$ for $x \in \mathbf{X}$ is a decision rule.

Given a decision rule δ,

$$R(\delta, \theta) = \iint L(a, \theta)\delta(da|x)P_\theta(dx) \tag{6.1}$$

is the *risk* of δ at θ. The function $\theta \to R(\delta, \theta)$ is called the *risk function* of δ. The risk function is used to compare decision rules with the goal being to find decision rules with "small" risk functions.

To introduce invariance into the decision problem, let G be a topological group which acts on the left of the three spaces $\mathbf{X}$, Θ and A. The word "space" here is being used as described in Section 1.1. The three group actions are not distinguished notationally, but the group action under consideration will be clear. For example, gx means the action of G on $\mathbf{X}$ evaluated at (g, x), while $g\theta$ means the action of G on Θ evaluated at (g, θ) and similarly for the action of G on A. It is emphasized that it is the group action which changes from space to space and not the group which changes. In other words, the group G remains fixed in our discussion and G acts in perhaps different ways on the different spaces. This is the reason for not adopting the more common notation used to describe invariant decision problems [for example, see Berger (1985), Chapter 6]. Given the ingredients of the decision problem listed above, here is the definition of a G-invariant decision problem when G acts on $\mathbf{X}$, Θ and A.

DEFINITION 6.2. The decision problem above is G-invariant if:

(i) The model $\{P_\theta|\theta \in \Theta\}$ is invariant, that is,

$$gP_\theta = P_{g\theta} \quad \text{for } g \in G, \theta \in \Theta.$$

(ii) The loss function L is invariant, that is

$$L(ga, g\theta) = L(a, \theta) \quad \text{for } g \in G, \theta \in \Theta, a \in A.$$

Now we turn to the invariance of decision rules. As in Example 2.9, the group G acts on $\mathbf{B}_3 \times \mathbf{X}$. Given a decision rule δ and $g \in G$, the decision rule $g\delta$ is defined by

$$(g\delta)(B|x) = \delta(g^{-1}B|g^{-1}x)$$

for $B \in \mathbf{B}_3$ and $x \in \mathbf{X}$.

DEFINITION 6.3. A decision rule δ is *invariant* if $g\delta = \delta$ for all $g \in G$.

When a decision rule δ is nonrandomized, say δ corresponds to a measurable function $x \to a_0(x)$, it is easy to show that δ is invariant iff the function a_0 is equivariant, that is, iff a_0 satisfies

$$a_0(gx) = ga_0(x)$$

for all $g \in G$ and $x \in \mathbf{X}$. This fact is used below without mention.

The primary focus of this chapter is to describe some techniques for finding "good" invariant decision rules. To this end we first describe some transformation formulas which are used later in the chapter.

THEOREM 6.1. *When the model $\{P_\theta | \theta \in \Theta\}$ is G-invariant, the formula*

$$\int f(gx) P_\theta(dx) = \int f(x) P_{g\theta}(dx) \tag{6.2}$$

holds for any function f which is integrable. For any decision rule δ, the formula

$$\int k(a)(g\delta)(da|x) = \int k(ga)\delta(da|g^{-1}x) \tag{6.3}$$

holds for any function k for which the integrals are well defined.

PROOF. Verify the formulas for the indicator functions of sets and then extend in the usual way. □

Throughout the remainder of this section and in the next section, it is assumed that we have a given invariant decision problem with the ingredients specified above. Here is a basic risk function identify for such problems.

THEOREM 6.2. *For any decision rule δ,*

$$R(g\delta, g\theta) = R(\delta, \theta) \tag{6.4}$$

for all $g \in G$ and $\theta \in \Theta$.

PROOF. We use (6.2) and (6.3) and calculate as

$$\begin{aligned} R(g\delta, \theta) &= \iint L(a, \theta)(g\delta)(da|x) P_\theta(dx) \\ &= \iint L(ga, \theta)\delta(da|g^{-1}x) P_\theta(dx) \\ &= \iint L(a, g^{-1}\theta)\delta(da|x) P_{g^{-1}\theta}(dx) = R(\delta, g^{-1}\theta). \end{aligned}$$ □

An important consequence of (6.4) is:

THEOREM 6.3. *If δ is an invariant decision rule, then*

$$R(\delta, g\theta) = R(\delta, \theta) \tag{6.5}$$

for all $g \in G$ and $\theta \in \Theta$.

PROOF. When δ is invariant, $g\delta = \delta$ so (6.4) yields (6.5). □

Equation (6.5) says that the risk function of an invariant decision rule δ is an invariant function of θ. In particular, when the action of G is transitive on Θ and δ is invariant,

$$R(\delta, \theta) = R(\delta, \theta_0), \qquad \theta \in \Theta,$$

for any fixed $\theta_0 \in \Theta$. In other words, when G is transitive on Θ, invariant decision rules have constant risk functions. In this situation, we expect a *best invariant decision rule* to exist because we can fix θ_0 and then try to minimize $R(\delta, \theta_0)$ over the class of all invariant decision rules. More precisely, δ_0 is called a *best invariant decision rule* if

$$R(\delta_0, \theta_0) \le R(\delta, \theta_0)$$

for all invariant rules δ. Of course, the choice of θ_0 is irrelevant because G is assumed to act transitively on Θ.

In the next section we describe one method of finding a best invariant decision rule when G is transitive on Θ. The method is originally due to Stein (unpublished) but the treatment here is rather different than I've seen in the literature. A related work is Zidek (1969).

6.2. Best invariant rules in the transitive case. Consider an invariant decision problem as described in the previous section. It is assumed throughout this section that G acts transitively on Θ. Hence all invariant decision rules have constant risk functions. Also assume that the Radon measure μ on $(\mathbf{X}, \mathbf{B}_1)$ dominates each P_θ in the statistical model $\{P_\theta | \theta \in \Theta\}$. The dominating measure μ is assumed to be relatively invariant with multiplier Δ^{-1} (Δ is the right hand modulus of G) as in Theorem 5.5. The densities

$$p(x|\theta) = \frac{dP_\theta}{d\mu}(x)$$

are, as usual, assumed to satisfy

$$p(x|\theta) = p(gx|g\theta)\Delta^{-1}(g) \tag{6.6}$$

for all x, θ and g.

The main assumption of this section, is that G acts properly on $\mathbf{X}$ (Definition 5.1). Therefore, the representation of J (the integral defined by μ) described in Theorem 5.5 holds. This representation involves the function T defined on $K(\mathbf{X})$ to $K(\mathbf{X}/G)$ by

$$T(f)(\pi(x)) = \int f(gx)\nu_r(dg), \tag{6.7}$$

where ν_r is right Haar measure on G and π is the natural projection from $\mathbf{X}$ to $\mathbf{X}/G$. Then Theorem 5.5 shows that

$$J(f) = J_1(T(f)) \tag{6.8}$$

for some integral J_1 on $K(\mathbf{X}/G)$. Of course (6.8) holds for all f which are μ-integrable.

We now apply (6.7) and (6.8) to the expression for the risk function of an invariant decision rule δ. Fix δ, fix $\theta \in \Theta$ and set

$$f_0(x) = \int_A L(a,\theta)\delta(da|x)p(x|\theta)$$

so that f_0 is nonnegative and

$$R(\delta,\theta) = \int f_0(x)\mu(dx) = J(f_0). \tag{6.9}$$

From (6.8), we have

$$R(\delta,\theta) = J_1(T(f_0)), \tag{6.10}$$

where

$$\begin{aligned} T(f_0)(\pi(x)) &= \int_G f_0(gx)\nu_r(dg) \\ &= \int_G\int_A L(a,\theta)\delta(da|gx)p(gx|\theta)\nu_r(dg). \end{aligned} \tag{6.11}$$

THEOREM 6.4. *The function* $T(f_0)$ *in* (6.11) *satisfies the equation*

$$T(f_0)(\pi(x)) = \int_G\int_A L(a,g\theta)\delta(da|x)p(x|g\theta)\nu_r(dg). \tag{6.12}$$

PROOF. Using (6.3) and (6.6) in (6.11) yields

$$T(f_0)(\pi(x)) = \int_G\int_A L(ga,\theta)\delta(da|x)p(x|g^{-1}\theta)\Delta(g)\nu_r(dg). \tag{6.13}$$

The invariance of L, the identity $\nu_l = \Delta\nu_r$, together with the fact that

$$\nu_l(dg^{-1}) = \nu_r(dg)$$

applied to (6.13) show that

$$\begin{aligned} T(f_0)(\pi(x)) &= \int_G\int_A L(a,g^{-1}\theta)\delta(da|x)p(x|g^{-1}\theta)\nu_l(dg) \\ &= \int_G\int_A L(a,g\theta)\delta(da|x)p(x|g\theta)\nu_r(dg). \end{aligned}$$

Thus (6.12) holds. □

Here is the description of how to find a best invariant decision rule. Define H on $A \times \mathbf{X}$ by

$$H(a,x) = \int_G L(a,g\theta)p(x|g\theta)\nu_r(dg). \tag{6.14}$$

Because G is transitive on Θ, the function H does not depend on $\theta \in \Theta$.

THEOREM 6.5. *Assume a measurable function a_0, defined on* **X** *with values in A, exists which satisfies*

$$(6.15)\qquad \begin{aligned} &\text{(i)}\quad H(a,x) \ge H(a_0(x),x) && \textit{for all } a, x,\\ &\text{(ii)}\quad a_0(gx) = g a_0(x) && \textit{for all } g, x. \end{aligned}$$

Then a_0 defines a best invariant decision rule δ_0 [that is, $\delta_0(\cdot|x)$ puts probability 1 at $a_0(x)$ and δ_0 is invariant by (ii)].

PROOF. Fix an invariant decision rule δ and fix $\theta \in \Theta$. Because G is transitive on Θ, it suffices to show that

$$R(\delta_0, \theta) \le R(\delta, \theta).$$

By definition of a_0 and δ_0, first observe that for any decision rule δ,

$$(6.16)\qquad \begin{aligned} &\int_G \int_A L(a, g\theta)\delta(da|x)p(x|g\theta)\nu_r(dg)\\ &\quad \ge \int L(a_0(x), g\theta)p(x|g\theta)\nu_r(dg)\\ &\quad = \int_G \int_A L(a, g\theta)\delta_0(da|x)p(x|g\theta)\nu_r(dg). \end{aligned}$$

Now, apply the integral J_1 to the first and last expressions in (6.16) and use (6.10) and (6.11) to yield $R(\delta_0, \theta) \le R(\delta, \theta)$. This completes the proof. □

The argument above provides a constructive method for finding a best invariant decision rule (under the stated assumptions). Namely, for each $x \in \mathbf{X}$, we minimize (over a) the function

$$(6.17)\qquad H(a, x) = \int L(a, g\theta)p(x|g\theta)\nu_r(dg)$$

to get the minimizer $a_0(x)$ (assuming it exists). It is an easy exercise to show that

$$H(ha, hx) = H(a, x) \quad \text{for } h \in G$$

and for all a and x. Thus the discussion of the orbit-by-orbit method given after Theorem 3.2 is valid. Under the regularity conditions of Theorem 3.2 applied to H, the resulting minimizer a_0 will satisfy (6.15)(ii). For simplicity of exposition, this invariance condition is simply assumed in the statement of Theorem 6.5.

There is a Bayesian interpretation of Theorem 6.5 which provides a partial answer to the question raised in the discussion after Theorem 3.2. Fix θ_0 in Θ and define the function τ on G to Θ by

$$\tau(g) = g\theta_0.$$

The map τ is onto because G is transitive. Now define the "induced" measure ξ on measurable subsets of Θ by

$$\xi(B) = \nu_r(\tau^{-1}(B)),$$

or equivalently

$$\int f(\theta)\xi(d\theta) = \int f(g\theta_0)\nu_r(dg),$$

for $f \in K(\Theta)$. When ξ is a well defined Radon measure [it may not be in some examples when $\xi(B) = +\infty$ for some compact B], then an easy calculation shows that ξ is relatively invariant with multiplier Δ^{-1}. Assuming ξ is well defined, the definition of ξ shows that

$$H(a, x) = \int L(a, \theta)p(x|\theta)\xi(d\theta). \tag{6.18}$$

But (6.18) is proportional to the posterior loss for taking action a when the prior (possibly improper) is ξ. Thus, a_0 can be interpreted as a formal Bayes rule for the prior distribution ξ. Note that ξ is a proper prior iff G is compact.

In some examples, the "natural" dominating measure for the probability measures of the model is not relatively invariant with multiplier Δ^{-1}, but is relatively invariant with some other multiplier. This situation was discussed earlier in the context of Theorem 5.6. To discuss this in the present context, assume that μ_0 is the dominating measure and μ_0 is relatively invariant with multiplier χ_0. From (5.16), the new dominating measure

$$\mu(dx) = \frac{1}{\phi(x)}\mu_0(dx)$$

is relatively invariant with multiplier Δ^{-1}. Here, ϕ is a strictly positive function on $\mathbf{X}$ which satisfies

$$\phi(gx) = \chi_0(g)\Delta(g)\phi(x) \quad \text{for } x \in X,\ g \in G.$$

The existence of such a ϕ was discussed in Chapter 5. Relative to this new dominating measure μ, the densities become

$$\tilde{p}(x|\theta) = p(x|\theta)\phi(x), \tag{6.19}$$

where $p(x|\theta)$ is the density of P_θ with respect to μ_0. Theorem 6.5 applied to $\tilde{p}(x|\theta)$ shows that a best invariant rule is formed by minimizing

$$\tilde{H}(a, x) = \int L(a, g\theta)\tilde{p}(x|g\theta)\nu_r(dg).$$

Using (6.19), we have

$$\tilde{H}(a, x) = \phi(x)H(a, x),$$

where

$$H(a, x) = \int L(a, g\theta)p(x|\theta)\nu_r(dg). \tag{6.20}$$

Because ϕ is strictly positive, minimizing $\tilde{H}$ is the same as minimizing H. The point is that as long as the dominating measure is relatively invariant with respect to some multiplier [and the densities satisfy (3.1) for that multiplier], then a best invariant rule is found by minimizing (6.20) for each $x \in \mathbf{X}$. Examples of this are given in the next section.

6.3. Examples of best equivariant estimators. Three examples in which best equivariant estimators are derived are presented here. The first concerns normal data with a known coefficient of variation. In this example the best equivariant estimator under normalized quadratic loss is different than the maximum likelihood estimator. The last two examples deal with the estimation of a covariance matrix based on data from a normal distribution. This case is quite interesting because the problem is invariant under two different groups and the resulting estimators are different for the loss functions considered.

EXAMPLE 6.1. Consider $X_1, \ldots, X_n$ which are iid $N(\theta, \theta^2)$ with $\theta > 0$. Thus the random vector X with coordinates $X_1, \ldots, X_n$ has the distribution

$$\mathscr{L}(X) = N(\theta e, \theta^2 I_n),$$

where e is the vector of 1's in R^n. With $\Theta = A = (0, \infty)$ and

$$L(a, \theta) = \frac{(a-\theta)^2}{\theta^2},$$

the ingredients of the decision problem are specified. Take G to be the multiplicative group $(0, \infty)$ so G acts on Θ and A in the obvious way. Clearly G is transitive on Θ. Further G acts on sample vectors $x \in R^n$ by coordinatewise multiplication. That the decision problem is invariant is easily checked.

The group G does not act properly on R^n, but G does act properly on the modified sample space $\mathbf{X} = R^n - \{0\}$. Thus the results of the previous section show that a best equivariant estimator is found by minimizing

$$H(a) = \int_0^\infty L(a, g\theta) f(x|g\theta)\frac{dg}{g},$$

where $f(\cdot|\theta)$ is the density of the data X with respect to Lebesgue measure on $\mathbf{X}$. Of course dg/g is a right Haar measure on G. Since $H(a)$ does not depend on the choice of θ, we take $\theta = 1$ for convenience. Thus, for each x, the function

$$H(a) = \int_0^\infty L(a, g) f(x|g)\frac{dg}{g}$$

needs to be minimized. With

$$q(\theta|x) = \frac{\theta^{-1} f(x|\theta)}{\int_0^\infty \theta^{-1} f(x|\theta)\, d\theta}.$$

$q(\cdot|x)$ can be viewed as a posterior density of θ given x obtained from the improper prior $d\theta/\theta$. Clearly, minimizing H is equivalent to minimizing

$$H_1(a) = \int_0^\infty L(a, \theta) q(\theta|x)\, d\theta$$

$$= \mathbf{E}\left[\frac{(a-\theta)^2}{\theta^2}\middle| X = x\right].$$

The minimum is easily shown to be

$$a_0(x) = \frac{\mathbf{E}[\theta^{-1}|X = x]}{\mathbf{E}[\theta^{-2}|X = x]},$$

which satisfies the equivariance condition (6.15)(ii). The best equivariant estimator a_0 is not known in closed form, but can be computed numerically. See Kariya (1984) for some further discussion.

For comparative purposes, the maximum likelihood estimator of θ is

$$\hat{\theta}(x) = \frac{-\bar{x} + \left[(\bar{x})^2 + 4\overline{x^2}\right]^{1/2}}{2},$$

where

$$\overline{x^2} = \frac{1}{n}\sum_1^n x_i^2.$$

Obviously $\hat{\theta}$ is equivariant, but it is not too hard to show that $\hat{\theta} \neq a_0$. This completes the first example. □

EXAMPLE 6.2. Consider iid p-dimensional random vectors $X_1, \ldots, X_n$ which have a multivariate normal distribution $N_p(0, \Sigma)$. The problem considered here is the estimation of the $p \times p$ covariance matrix Σ which is assumed to be nonsingular, but otherwise unknown. Further, it is assumed that $n \geq p$ so that the sufficient statistic

$$S = \sum_{i=1}^{n} X_i X_i'$$

is positive definite with probability 1. Without loss of generality, estimators of Σ are functions of S. Thus the sample space $\mathbf{X}$, the parameter space Θ and the action space A are taken to be the set of $p \times p$ positive definite matrices. Obviously S has a Wishart distribution $W(\Sigma, p, n)$ and the maximum likelihood estimator of Σ is

$$\hat{\Sigma} = n^{-1}S.$$

The group Gl_p acts on $\mathbf{X}$, Θ and A in the obvious way:

$$S \to gSg',$$
$$\Sigma \to g\Sigma g',$$
$$a \to gag',$$

for $g \in \mathrm{Gl}_p$. The model for S is invariant under this group action.

Now, consider a loss function L which is invariant, that is, L satisfies

$$L(gag', g\Sigma g') = L(a, \Sigma) \tag{6.21}$$

for all g, a and Σ. A standard invariance argument shows that L satisfies (6.21) iff L can be written as a function of the eigenvalues of $a\Sigma^{-1}$, say $\lambda_1 \geq \cdots \geq$

$\lambda_p > 0$. Two interesting examples are

$$
\begin{aligned}
L_1(a, \Sigma) &= \operatorname{tr} \Sigma^{-1}(a - \Sigma)\Sigma^{-1}(a - \Sigma) = \operatorname{tr}\left(\Sigma^{-1/2} a \Sigma^{-1/2} - I_p\right)^2 \\
&= \sum_{i=1}^{p} (\lambda_i - 1)^2
\end{aligned} \tag{6.22}
$$

and

$$
\begin{aligned}
L_2(a, \Sigma) &= \operatorname{tr} a\Sigma^{-1} - \log\det(a\Sigma^{-1}) - p \\
&= \sum_{i=1}^{p} (\lambda_i - \log \lambda_i - 1).
\end{aligned} \tag{6.23}
$$

The loss function L_1 was used by Selliah (1964) in his Stanford thesis [also see Olkin and Selliah (1977)]. Stein (1956) introduced L_2 in his study of covariance estimation [also see James and Stein (1960)].

Because Gl_p acts transitively on Θ, a best equivariant (nonrandomized) estimator should exist. Rather than use the results of Section 6.2, it is a bit easier for this particular example to proceed as follows. An estimator $\tau(S)$ is equivariant iff

$$\tau(gSg') = g\tau(S)g' \tag{6.24}$$

for all S and $g \in \mathrm{Gl}_p$. Picking $g = S^{-1/2}$, (6.24) implies that

$$\tau(S) = S^{1/2}\tau(I)S^{1/2}. \tag{6.25}$$

But, for any $\gamma \in O_p$, (6.24) implies

$$\tau(I) = \tau(\gamma I \gamma') = \gamma\tau(I)\gamma', \tag{6.26}$$

which implies that the matrix $\tau(I)$ must be a multiple of the identity matrix, say $\tau(I) = \alpha I$. Combining this with (6.25) shows that an estimator is equivariant iff

$$\tau(S) = \alpha S \tag{6.27}$$

for some real α. Thus, to find a best equivariant estimator, the constant α is selected to minimize the risk at some (any) fixed point in Θ. For L_1, α is to be selected to minimize the risk

$$
\begin{aligned}
R_1(\alpha S, I) &= \mathbf{E}_I \operatorname{tr}(\alpha S - I)^2 \\
&= \alpha^2 \mathbf{E}_I \operatorname{tr} S^2 - 2\alpha \mathbf{E}_I \operatorname{tr} S + p.
\end{aligned}
$$

The minimizer is

$$\alpha_1 = \frac{\mathbf{E}_I \operatorname{tr} S}{\mathbf{E}_I \operatorname{tr} S^2} = \frac{np}{(n^2 - 2n)p + n(p-1)} = \frac{1}{n + 2 + (p-1)/p}.$$

Using L_2, a similar calculation shows that the minimizing α is

$$\alpha_2 = \frac{1}{n}.$$

Hence for L_2, the maximum likelihood estimator is a best equivariant estimator.

Because the group of $p \times p$ lower triangular matrices with positive diagonal elements G_T^+ is a subgroup of Gl_p, the above estimation problem (with loss functions L_1 or L_2) is also invariant under G_T^+. Since G_T^+ acts transitively on Θ, a best equivariant estimator should exist for this problem. In this case an estimator $\tau(S)$ is equivariant iff

$$\tau(gSg') = g\tau(S)g', \qquad g \in G_T^+, \tag{6.28}$$

for each $S \in \mathbf{X}$. Standard arguments show that (6.28) holds iff

$$\tau(S) = TAT', \tag{6.29}$$

where T is the unique element in G_T^+ such that $S = TT'$ and A is a fixed symmetric matrix. Thus finding a best equivariant (under G_T^+) estimator involves finding A to minimize the risk at a fixed point in Θ. This of course depends on the loss function.

For loss function L_1, A is chosen to minimize

$$R_1 = \mathbf{E}_I \operatorname{tr}(TAT' - I)^2.$$

The minimizing A is a diagonal matrix, but the minimizer is not known explicitly. For the equations determining the minimizer see Selliah (1964) and Olkin and Selliah (1977).

When the loss function is L_2, A is chosen to minimize

$$\begin{aligned} R_2 &= \mathbf{E}_I[\operatorname{tr} TAT' - \log\det(TAT') - p] \\ &= \mathbf{E}_I \operatorname{tr} T'TA - \log\det(A) + c_0, \end{aligned}$$

where c_0 is a constant. Thus, it suffices to minimize

$$\tilde{R}_2 = \operatorname{tr} \mathbf{E}(T'T)A - \log\det(A) = \sum_{i=1}^{p} d_{ii}a_{ii} - \log\det(A),$$

where $a_{ii}, \ldots, a_{pp}$ are the diagonal elements of A and

$$d_{ii} = n + p - 2i + 1, \qquad i = 1, \ldots, p.$$

The minimum is achieved at

$$A_0 = D^{-1},$$

where D is diagonal with diagonal elements $d_{11}, \ldots, d_{pp}$. Therefore the best equivariant estimator in this case is

$$\tau_2(S) = TD^{-1}T'$$

as was established by Stein [see James and Stein (1960)]. It also follows from results in Kiefer (1957) that this estimator τ_2 is minimax [because the group G_T^+ is solvable; see Bondar and Milnes (1981) for a survey], but the maximum likelihood estimator $\hat{\Sigma}$ is not minimax when L_2 is the loss function. For some related results, see Eaton and Olkin (1987). This completes Example 6.2. □

It should be noted that the results established in Section 6.2 were not used to find the best equivariant estimators in Example 6.2. Rather, the procedure was to first characterize the functional form of the equivariant estimators and then minimize the risk (at a fixed point in Θ) over the class of equivariant estimators. This was feasible because the functional form was rather simple for the two cases considered. In the next example, which also concerns covariance estimation, this procedure seems not to be feasible because the class of equivariant estimators is rather large. For this reason, the method described in Section 6.2 is used in the following example.

EXAMPLE 6.3. As in Example 6.2, assume that data S which is $W(\Sigma, p, n)$ with $n \geq p$, is available. Partition Σ as

$$\Sigma = \begin{pmatrix} \Sigma_{11} & \Sigma_{12} \\ \Sigma_{21} & \Sigma_{22} \end{pmatrix},$$

where Σ_{11} is $q \times q$ and Σ_{22} is $r \times r$, so $p = q + r$. Consider "extra" data $X_1, \ldots, X_m$ which consists of iid random vectors in R^q such that each X_i has a $N(0, \Sigma_{11})$ distribution. The problem is to estimate Σ using S and the data $X_1, \ldots, X_m$. Note that (S, V) is a sufficient statistic where

$$V = \sum_1^m X_i X_i'.$$

To describe the invariance of this problem, let G be the subgroup of Gl_p whose elements g have the form

$$g = \begin{pmatrix} g_{11} & 0 \\ g_{21} & g_{22} \end{pmatrix}$$

with $g_{11} \in \mathrm{Gl}_q$ and $g_{22} \in \mathrm{Gl}_r$. The model is easily shown to be invariant under the group actions

$$(S, V) \to (gSg', g_{11}Vg_{11}'),$$
$$\Sigma \to g\Sigma g'.$$

This invariance together with the orbit-by-orbit method described in Theorem 3.2 can be used to derive the maximum likelihood estimator for Σ. The details are not given here [see Eaton (1970) for one derivation of $\hat{\Sigma}$].

Because the group G_T^+ of $p \times p$ lower triangular matrices with positive diagonal elements is a subgroup of G, the model for the data is G_T^+ invariant. Also, G_T^+ acts transitively on Θ so a best equivariant estimator should exist. In the remainder of this example we focus on finding a best equivariant (under G_T^+) estimator when the loss function is L_2 given in Example 6.2. First, we characterize the equivariant estimators, that is, those estimators τ which satisfy

$$\tau(gSg', g_{11}Vg_{11}') = g\tau(S, V)g' \tag{6.30}$$

for $g \in G_T^+$ and for sample points (S, V) for which S is positive definite and V is nonnegative definite. Let T be the unique element in G_T^+ satisfying $S = TT'$. Picking $g = T^{-1}$ in (6.30) shows that τ satisfying (6.30) must satisfy

$$\tau(TT', V) = T\tau\big(I, T_{11}^{-1}V(T_{11}^{-1})'\big)T' = T\tau_0\big(T_{11}^{-1}V(T_{11}^{-1})'\big)T',$$

where τ_0 maps $q \times q$ nonnegative definite matrices into $p \times p$ positive definite matrices (since τ is assumed to take values in Θ). Conversely, if τ is given by

$$\tau(TT', V) = T\tau_0\big(T_{11}^{-1}V(T_{11}^{-1})'\big)T', \tag{6.31}$$

then it is easy to show that τ satisfies (6.30). Hence, all equivariant estimators have the form (6.31) where τ_0 is "arbitrary." Thus, to find a best equivariant estimator, we must minimize the risk over all τ_0's:

$$R_2 = \mathbf{E}_I L_2\big[T\tau_0\big(T_{11}^{-1}V(T_{11}^{-1})'\big)T, I\big].$$

A straightforward approach to this problem, such as that used in Example 6.2, seems very difficult to carry out. For this reason, the method described in Section 6.2 is used.

To show that the method of Section 6.2 is applicable, first note that the sample space $\mathbf{X} \times (R^q)^m$ is acted on by G_T^+ via

$$(S, x_1, \dots, x_m) \to (gSg', g_{11}x_1, \dots, g_{11}x_m),$$

where $S \in \mathbf{X}$ (as in Example 6.2), each x_i is in R^q and $g \in G_T^+$. Lebesgue measure on $\mathbf{X} \times (R^q)^m$ is relatively invariant under the group action and the action is proper. Thus a best equivariant estimator is found by minimizing [from (6.20)]

$$H(a) = \int L_2(a, g\theta_0)p(z|g\theta_0)\nu_r(dg), \tag{6.32}$$

where θ_0 is a fixed point in Θ, $z \in \mathbf{X} \times (R^q)^m$ is a sample point and ν_r is a right-invariant measure on G_T^+. For convenience, pick $\theta_0 = I_p \in \Theta$. Substituting the explicit form of the density p in (6.32), the function which needs to be minimized is

$$\begin{aligned} H(a) = &\int L_2(a, gg')\big|(gg')^{-1}S\big|^{n/2}|g_{11}g_{11}'|^{-m/2} \\ &\times \exp\big[-\tfrac{1}{2}\operatorname{tr}(gg')^{-1}S - \tfrac{1}{2}\operatorname{tr}(g_{11}g_{11}')^{-1}V\big]\nu_r(dg). \end{aligned} \tag{6.33}$$

The orbit-by-orbit method shows it is sufficient to minimize (6.33) when $S = I_p$. Setting $S = I_p$ and making the change of variable $g \to g^{-1}$, the function we want to minimize is

$$H_1(a) = \int L_2(a, gg')|g'g|^{n/2}|g_{11}'g_{11}|^{m/2}\exp\big[-\tfrac{1}{2}\operatorname{tr} g'g - \tfrac{1}{2}\operatorname{tr} g_{11}'g_{11}V\big]\nu_l(dg),$$

where ν_l is a left-invariant measure on G_T^+. The details of this minimization are given in Eaton (1970) and are not described further here. The best equivariant estimator obtained from this minimization is as follows. First write the data (S, V) as

$$S = TT', \qquad U = T_{11}^{-1}V(T_{11}^{-1})',$$

where $T \in G_T^+$. Further, write

$$I_q + U = WW', \tag{6.34}$$

where W is the unique $q \times q$ lower triangular matrix with positive diagonals satisfying (6.34). The best equivariant estimator (for loss function L_2) can be written

$$\tau(S) = T\tau_0(U)T',$$

where

$$\tau_0(U) = \begin{bmatrix} \left[(W^{-1})'DW^{-1} + (p-q)I\right]^{-1} & 0 \\ 0 & E^{-1} \end{bmatrix}$$

and D: $q \times q$ and E: $r \times r$ are fixed diagonal matrices with diagonal elements

$$d_{ii} = m + n + q - 2i + 1, \qquad i = 1, \ldots, q,$$

and

$$e_{ii} = n + p - 2q - 2i + 1, \qquad i = 1, \ldots, r.$$

This estimator is minimax because of results in Kiefer (1957). □

6.4. Invariant testing examples. Here we consider an invariant testing problem where the representation theorem, Theorem 5.9, is applicable. On a space $\mathbf{X}$ which is acted on *properly* by a locally compact group G, assume that a Radon measure μ is relatively invariant with multiplier χ. A family of densities with respect to μ, say $\{p(\cdot|\theta)|\theta \in \Theta\}$, is given. Further, the group G acts on Θ and the basic invariance condition

$$p(x|\theta) = p(gx|g\theta)\chi(g) \quad \text{for } x, \theta, g$$

is assumed to hold. Thus the probability model on $\mathbf{X}$ determined by the family of densities is invariant.

Let Θ_0 and Θ_1 be G invariant subsets of Θ and consider testing

$$H_0\colon \theta \in \Theta_0 \quad \text{versus} \quad H_1\colon \theta \in \Theta_1.$$

This testing problem is, according to the discussion given in Section 3.2, invariant. In what follows, attention is restricted to invariant test functions ϕ. Thus, if $\tau(X)$ is a maximal invariant, an invariant test function ϕ can be written as

$$\phi(X) = \psi(\tau(X)).$$

Now, we add the final assumption that *G acts transitively on both Θ_0 and Θ_1.* Under this assumption (plus those above), a most powerful level α invariant test is given below. To describe this test, first observe that the power function of any invariant test, say

$$\beta_\phi(\theta) = \mathbf{E}_\theta \phi(X),$$

is an invariant function of θ. Because G is transitive on Θ_0 and Θ_1, this power function takes on only two values, namely,

$$a_i = \beta_\phi(\theta), \qquad \theta \in \Theta_i, \qquad i = 0, 1.$$

Thus, a_0 is the level of ϕ and a_1 is the power of ϕ. But, for fixed a_0, the Neyman–Pearson lemma tells us that the most powerful test rejects for large values of the ratio

$$r(t) = \frac{q_1(t)}{q_0(t)},$$

where q_i is the density of $\tau(X)$ when X has density $p(\cdot|\theta)$, $\theta \in \Theta_i$ for $i = 0, 1$. Of course, q_i does not depend on $\theta \in \Theta_i$ because G is transitive on Θ_i. Note that densities always exist in the present situation because $\tau(X)$ has only two possible distributions—one under H_0 and one under H_1. However, Theorem 5.9 tells us how to compute the ratio $r(t)$. In the notation of this section, fix $\theta_i \in \Theta_i$, $i = 0, 1$. Then according to (5.20) [using the maximal invariant $\pi(x)$],

$$r(\pi(x)) = \frac{\int p(gx|\theta_1)\chi(g)\nu_l(dg)}{\int p(gx|\theta_0)\chi(g)\nu_l(dg)}, \tag{6.35}$$

where χ is the multiplier specified in the model and ν_l is a left-invariant measure on G. Naturally the denominator is assumed to be positive. Also, the expression (6.35) does not depend on the choices of θ_0 and θ_1 because of the transitivity assumption. The above discussion implies that the most powerful level α-invariant test rejects for large values of $r(\pi(x))$ given in (6.35).

Before turning to some examples, it is useful to contrast the hypothesis testing problem considered here with the estimation problem of the last section. It was assumed that G acted transitively in the estimation problem, and this assumption implied that equivariant estimators had constant risk. Thus the risk of decision rule in the estimation problem is a single number, because there is only one orbit in the parameter space. In the testing problem above, the parameter space is $\Theta_0 \cup \Theta_1$ and there are two orbits (assuming $\Theta_0 \neq \Theta_1$). Thus, the risk function of an invariant test is determined by two numbers. In this situation, the Neyman–Pearson lemma tells us what the good invariant decision rules are.

EXAMPLE 6.4. This example concerns Hotelling's T^2 test for certain types of nonnormal data and comes from Kariya (1981). Consider a random matrix X: $n \times p$ which has a density with respect to Lebesgue measure, on np dimensional space given by

$$p(x|\theta) = |\Sigma|^{-n/2} f\left[\operatorname{tr}(x - e\mu')\Sigma^{-1}(x - e\mu')'\right]. \tag{6.36}$$

Here x is $n \times p$, e is the vector of 1's in R^n, μ is an unknown vector in R^p, Σ is a $p \times p$ positive definite matrix, tr denotes the trace and f is some nonnegative function defined on $[0, \infty)$ which satisfies

$$\int f\left[\operatorname{tr} x'x\right] dx = 1.$$

The parameter θ stands for the pair (μ, Σ). In the case that

$$f(z) = (\sqrt{2\pi})^{-np} \exp\left[-\tfrac{1}{2}z\right], \qquad z \in [0, \infty],$$

then of course the rows of X are iid $N_p(\mu, \Sigma)$.

Here is the classical hypothesis testing problem which Hotelling's T^2 test solves for normal data:

$$H_0: \mu = 0 \quad \text{versus} \quad H_1: \mu \neq 0.$$

The matrix Σ is unrestricted under both H_0 and H_1. With a rotation of coordinates and some relabelling, we can (and do) replace the vector e in (6.36) with the vector

$$\varepsilon_1 = \begin{pmatrix} 1 \\ 0 \\ \vdots \\ 0 \end{pmatrix} \in R^n.$$

This relabelling results in some notational simplification. It is assumed that $n \geq p + 1$ and matrices x: $n \times p$ are partitioned as

$$x = \begin{pmatrix} x_1 \\ x_2 \end{pmatrix},$$

where x_1 is $1 \times p$ and x_2 is $(n-1) \times p$. The sample space for this problem is taken to be the set of x's such that x_2: $(n-1) \times p$ has rank p.

To describe a group under which this problem is invariant, let G be the product group $G_0 \times \mathrm{Gl}_p$ where G_0 is the subgroup of O_n whose elements have the form

$$g = \begin{pmatrix} 1 & 0 \\ 0 & \gamma \end{pmatrix}, \qquad \gamma \in O_{n-1}.$$

Given $(g, a) \in G_0 \times \mathrm{Gl}_p$, the group action on $\mathbf{X}$ is

$$x \to gxa'$$

and the group action on the parameter $\theta = (\mu, \Sigma)$ is

$$(\mu, \Sigma) \to (a\mu, a\Sigma a').$$

Note that Lebesgue measure is relatively invariant with multiplier

$$\chi(g, a) = |\det(a)|^n.$$

Routine calculations show that a maximal invariant function on $\mathbf{X}$ is

$$T^2 = X_1(X_2'X_2)^{-1}X_1',$$

where

$$X = \begin{pmatrix} X_1 \\ X_2 \end{pmatrix}.$$

Further, a maximal invariant function on the parameter space is

$$\delta = \mu'\Sigma^{-1}\mu.$$

Now, fix $\delta_1 > 0$ and consider

$$\Theta_0 = \{(\mu, \Sigma) | \delta = 0\}$$

and

$$\Theta_1 = \{(\mu, \Sigma) | \delta = \delta_1\}.$$

The group G acts transitively on Θ_0 and Θ_1 and G acts properly on $\mathbf{X}$. Thus, a best invariant level α test for testing the null hypothesis

$$H_0: \theta \in \Theta_0 \quad \text{versus} \quad \tilde{H}: \theta \in \Theta_1$$

is found by calculating r in (6.35). Clearly the original null hypothesis is equivalent to $\theta \in H_0$, but the original alternative is H_1: $\delta > 0$. Our first goal is to find a best invariant test for H_0 versus $\tilde{H}_1$. To this end, take $\theta_0 \in \Theta$ to be

$$\theta_0 = (0, I_p)$$

and θ_1 to be

$$\theta_1 = \left(\sqrt{\delta_1}\,\xi, I_p\right),$$

where

$$\xi = \begin{pmatrix} 1 \\ 0 \\ \vdots \\ 0 \end{pmatrix} \in R^p.$$

A left-invariant measure on G is

$$\nu_l(dh) = \nu(dg)\frac{da}{|\det(a)|^p},$$

where ν is the invariant probability measure on G_0 and da is Lebesgue measure on Gl_p. Substituting these and (6.36) into (6.35), and changing variables shows that

$$r(v) = \frac{\int f\left[\operatorname{tr} a'a + \delta_1 - 2a_{11}v\delta_1^{1/2}\right]|a'a|^{(n-p)/2}\,da}{\int f\left[\operatorname{tr} a'a\right]|a'a|^{(n-p)/2}\,da}, \tag{6.37}$$

where both of the integrals are over Gl_p, a_{11} is the (1, 1) element of a and

$$v = \frac{T^2}{\left(1 + T^2\right)^{1/2}}.$$

Obviously v is an increasing function of T^2 and v is also a maximal invariant statistic since it is a one-to-one function of T^2.

PROPOSITION 6.1. *Assume f is a convex function on $[0, \infty)$. Then the best invariant level α test of H_0 versus $\tilde{H}_1$ rejects for large values of T^2. Further, the null distribution of T^2 is that when*

$$f(z) = \left(\sqrt{2\pi}\right)^{-np}\exp\left[-\tfrac{1}{2}z\right].$$

PROOF. Because Lebesgue measure on Gl_p is invariant under the transformation $a \to -a$, it follows that

$$r(v) = r(-v), \qquad v \geq 0,$$

where r is given in (6.37). The convexity of f implies that for $\frac{1}{2} \leq \beta \leq 1$ and

$v \geq 0$, we have

$$r((2\beta - 1)v) = r(\beta v + (1 - \beta)(-v))$$
$$\leq \beta r(v) + (1 - \beta)r(-v) = r(v).$$

Hence $r(v)$ is nondecreasing in v so rejecting for large values of r is equivalent to rejecting for large values of v, which is, in turn, equivalent to rejecting for large values of T^2. The first part of the proposition is proved.

That the null distribution of T^2 does not depend on the particular f in (6.36) is a consequence of the null robustness results described in Section 4.3. □

An immediate consequence of Proposition 6.1 is that the test which rejects for large values of T^2 is a best invariant test (of its level) for testing H_0 versus H_1. This follows because the best invariant test of H_0 versus $\tilde{H}_1$ did not depend on the particular alternative $\tilde{H}_1$. Thus, as long as f is convex, Hotelling's T^2 test as a best invariant test of H_0 versus H_1 and the null distribution of T^2 is known for each f. This completes our discussion of Example 6.4. □

EXAMPLE 6.5. Here we briefly discuss a relatively smooth example where a best invariant test exists (Wijsman's theorem applies) and this test is different from the likelihood ratio test. This example is essentially due to Stein and was originally constructed to show that the Hunt–Stein theorem is not valid for the group Gl_p, $p \geq 2$. See Lehmann [(1959), Problem 10, page 344] for a related result.

Consider two independent Wishart matrices S_1 with

$$\mathcal{L}(S_1) = W(\Sigma, p, n)$$

and S_2 with

$$\mathcal{L}(S_2) = W(c\Sigma, p, n).$$

It is assumed $n \geq p \geq 2$, Σ is an unknown $p \times p$ positive definite matrix and the real constant c is positive. The problem is to test

$$H_0\colon c = 1 \quad \text{versus} \quad H_1\colon c = 2$$

so Σ is a nuisance parameter. In our previous notation,

$$\Theta_0 = \{(c, \Sigma) | c = 1\}$$

and

$$\Theta_1 = \{(c, \Sigma) | c = 2\}.$$

It is easily verified that this problem is invariant under Gl_p acting on S_i by

$$S_i \to g S_i g', \qquad g \in \mathrm{Gl}_p$$

and on Σ by

$$\Sigma \to g \Sigma g', \qquad g \in \mathrm{Gl}_p.$$

Clearly Gl_p is transitive on Θ_0 and Θ_1. The other conditions necessary to apply the argument at the beginning of this section are easily checked. Fix a level α in

$(0, 1)$. The ratio in (6.35) defines a best level α Gl_p-invariant test, say ϕ_0. Note that the likelihood ratio test is Gl_p-invariant so can be no better than ϕ_0.

Now, G_T^+ is a subgroup of Gl_p and so the testing problem above is also invariant under G_T^+. Also G_T^+ acts transitively on Θ_0 and Θ_1. Again the conditions necessary to apply the argument at the beginning of this section can be verified. Thus again the ratio in (6.35) computed using G_T^+ defines a best G_T^+-invariant test, say ϕ_1. The test ϕ_1 is at least as good as ϕ_0 because ϕ_0 is G_T^+-invariant. In fact ϕ_1 is a better test than ϕ_0 and hence ϕ_1 dominates the likelihood ratio test.

The point of this example is that there is a bit "too much" invariance in the problem above. Fully invariant procedures such as the likelihood ratio test can be improved upon by simply requiring less invariance. This completes Example 6.5. □

Finally we mention the book by Kariya and Sinha (1988) which contains material on null and nonnull robustness as well as further applications of Wijsman's theorem.

CHAPTER 7

Random Orthogonal Matrices

Orthogonal matrices, both fixed and random, play an important role in much of statistics, especially in multivariate analysis. Connections between the orthogonal group O_n and the multivariate normal distribution are explored in James (1954) and in Wijsman (1957), as well as in many texts on multivariate analysis. In this chapter, invariance arguments are used to derive the density of a subblock of a uniformly distributed element of O_n. This result is used to describe an upper bound on the rate at which one has convergence (as $n \to \infty$) to the multivariate normal distribution.

7.1. Generating a random orthogonal. Throughout this chapter, O_n denotes the group of $n \times n$ real orthogonal matrices. By the uniform distribution on the compact group O_n, we mean the unique left- (and right-) invariant probability measure. It is possible to represent this distribution using differential forms, but the approach taken here is to represent things in terms of random matrices.

Consider a random matrix X: $n \times q$ with $q \leq n$ and assume

$$\mathscr{L}(X) = N(0, I_n \otimes I_q).$$

Thus, the elements of X are iid $N(0,1)$ random variables. It is well known that X has rank q with probability 1 [for example, see Eaton (1983), Chapter 7]. Thus the random matrix

$$\Gamma_1 = X(X'X)^{-1/2}$$

is well defined. Since $\Gamma_1'\Gamma_1 = I_q$, Γ_1 is a random element of $F_{q,n}$ introduced in Example 2.3. Since O_n acts transitively on $F_{q,n}$, there exists a unique invariant probability measure on $F_{q,n}$, say ν. The following result shows that Γ_1 has distribution ν.

PROPOSITION 7.1. *The random matrix Γ_1 has the uniform distribution on $F_{q,n}$.*

PROOF. From the uniqueness of ν, it suffices to show that

$$\mathscr{L}(\Gamma_1) = \mathscr{L}(g\Gamma_1), \qquad g \in O_n.$$

But, it is clear that

$$\mathscr{L}(X) = \mathscr{L}(gX), \qquad g \in O_n.$$

Thus,

$$\begin{aligned}\mathscr{L}(g\Gamma_1) &= \mathscr{L}\big(gX(X'X)^{-1/2}\big)\\ &= \mathscr{L}\big(gX((gX)'gX)^{-1/2}\big)\\ &= \mathscr{L}\big(X(X'X)^{-1/2}\big) = \mathscr{L}(\Gamma_1).\end{aligned}$$

□

When $q = 1$, Γ_1 has the uniform distribution on the unit sphere in R^n and when $q = n$, Γ_1 has the uniform distribution on $O_n = F_{n,n}$. The two properties of X which lead to Proposition 7.1 are:

(i) X has rank q a.s.
(ii) $\mathscr{L}(X) = \mathscr{L}(gX)$, $g \in O_n$.

Any random matrix X satisfying (i) and (ii) yields a Γ_1 which is uniform on $F_{q,n}$.

Now, partition X as

$$X = \begin{pmatrix} Y \\ Z \end{pmatrix}$$

with Y: $p \times q$. Then

$$\Gamma_1 = \begin{pmatrix} Y \\ Z \end{pmatrix}(Y'Y + Z'Z)^{-1/2}$$

so that

$$\Delta = Y(Y'Y + Z'Z)^{-1/2} \tag{7.1}$$

is the $p \times q$ upper block of Γ_1. When $p + q \le n$, Khatri (1970) derived the density of Δ using the invariant differential on O_n. Here we derive the density of Δ using an invariance argument.

The following should be noted. If Γ is uniform on O_n and Γ is partitioned as

$$\Gamma = (\Gamma_1 \Gamma_2)$$

with Γ_1: $n \times q$ and Γ_2: $n \times (n - q)$, then Γ_1 has the uniform distribution on $F_{q,n}$. This follows from the observation that for $g \in O_n$,

$$\mathscr{L}(\Gamma) = \mathscr{L}(g\Gamma) = \mathscr{L}(g(\Gamma_1\Gamma_2)) = \mathscr{L}((g\Gamma_1 g\Gamma_2))$$

so that marginally

$$\mathscr{L}(\Gamma_1) = \mathscr{L}(g\Gamma_1).$$

This invariance characterizes $\mathscr{L}(\Gamma_1)$. Therefore, the matrix Δ: $p \times q$ can be thought of as the $p \times q$ upper left block of the random orthogonal matrix Γ.

7.2. The density of a block. Before turning to the density of Δ, we review a few basic facts about the multivariate beta distribution. Consider two independent Wishart matrices S_i, $i = 1, 2$, with

$$\mathscr{L}(S_i) = W(I_q, q, n_i), \qquad i = 1, 2.$$

That is, S_i: $q \times q$ has a Wishart distribution with n_i degrees of freedom and scale matrix I_q. When $n_1 + n_2 \geq q$, then $S_1 + S_2$ is positive definite with probability 1 and the random matrix

$$B = (S_1 + S_2)^{-1/2} S_1 (S_1 + S_2)^{-1/2}$$

is well defined. The matrix B has, by definition, a multivariate beta distribution which is written

$$\mathscr{L}(B) = \mathbf{B}(n_1, n_2; I_q).$$

This notation is from Dawid (1981). Since

$$I_q - B = (S_1 + S_2)^{-1/2} S_2 (S_1 + S_2)^{-1/2},$$

it follows that B can have a density with respect to Lebesgue measure [on the set of symmetric B's with all eigenvalues in $(0, 1)$] iff $n_i \geq q$, $i = 1, 2$. In this case, the density of B is

$$\frac{\omega(n_1, q)\omega(n_2, q)}{\omega(n_1 + n_2, q)} |B|^{(n_1 - q - 1)/2} |I_q - B|^{(n_2 - q - 1)/2}, \tag{7.2}$$

where $\omega(\cdot, \cdot)$ is the Wishart constant [Eaton (1983), page 175]. This and related results can be found in Olkin and Rubin (1964), Mitra (1970) and Khatri (1970).

In what follows, we treat the case $q \leq p$ in the discussion of

$$\Delta = Y(Y'Y + Z'Z)^{-1/2},$$

where

$$\mathscr{L}(X) = \mathscr{L}\left(\begin{pmatrix} Y \\ Z \end{pmatrix}\right) = N(0, I_n \otimes I_q).$$

The case $p \leq q$ is treated by taking transposes.

Proposition 7.2. *When $q \leq p$ and $p + q \leq n$, the random matrix $\Delta'\Delta$ has a $\mathbf{B}(p, n - p; I_q)$ distribution. Thus $\Delta'\Delta$ has a density (with respect to Lebesgue measure) given by*

$$f_0(x) = C_0 |x|^{(p - q - 1)/2} |I_q - x|^{(n - p - q - 1)/2} I_0(x), \tag{7.3}$$

where I_0 is the indicator function of the $q \times q$ symmetric matrices all of whose eigenvalues are in $(0, 1)$ and the constant C_0 is

$$C_0 = \omega(p, q)\omega(n - p, q)/\omega(n, q).$$

Proof. Using the normal representation for Δ, we have

$$\begin{aligned}\mathscr{L}(\Delta'\Delta) &= \mathscr{L}\left((Y'Y + Z'Z)^{-1/2} Y'Y (Y'Y + Z'Z)^{-1/2}\right) \\ &= \mathscr{L}\left((S_1 + S_2)^{-1/2} S_1 (S_1 + S_2)^{-1/2}\right).\end{aligned}$$

Since S_1 is $W(I_q, q, p)$ and S_2 is $W(I_q, q, n-p)$, the first assertion follows. The expression for the density of $\Delta'\Delta$ follows immediately from (7.2). □

PROPOSITION 7.3. *For $q \le p$ and $p + q \le n$, the density of Δ is given by*

$$f_1(y) = C_1 |I_q - y'y|^{(n-p-q-1)/2} I_0(y'y), \tag{7.4}$$

where I_0 is given in Proposition 5.2 *and the constant C_1 is*

$$C_1 = (\sqrt{2\pi})^{-pq} \omega(n-p, q)/\omega(n, q). \tag{7.5}$$

PROOF. Let ψ: $p \times q$ have f_1 as a density. Since $f_1(gy) = f_1(y)$ for $g \in O_p$, it is clear that

$$\mathscr{L}(\psi) = \mathscr{L}(g\psi), \qquad g \in O_p.$$

Because $\psi'\psi$ is a maximal invariant under the group action

$$\psi \to g\psi, \qquad g \in O_p,$$

the results of Example 5.2 show that $\psi'\psi$ has the density f_0 given in (7.3). Therefore,

$$\mathscr{L}(\psi'\psi) = \mathscr{L}(\Delta'\Delta).$$

However, we know that

$$\mathscr{L}(g\Delta) = \mathscr{L}(\Delta), \qquad g \in O_p.$$

The results of Proposition 7.4 below imply that $\mathscr{L}(\psi) = \mathscr{L}(\Delta)$ so that Δ has density (7.3). □

Essentially, the argument used above to conclude that $\mathscr{L}(\psi) = \mathscr{L}(\Delta)$ consists of two parts:

(i) Both ψ and Δ have O_p-invariant distributions.
(ii) The distribution of the maximal invariant under the group action is the same for both ψ and Δ.

The result below shows that the argument is, in fact, a general argument and not specific to the case at hand. To describe the general situation, suppose that the compact group G acts measurably on measurable space $\mathbf{Y}$ and suppose that P_1 and P_2 are two G-invariant probability measures on $\mathbf{Y}$. Let τ: $\mathbf{Y} \to \mathbf{X}$ be a maximal invariant function.

PROPOSITION 7.4. *Let $Y_i \in \mathbf{Y}$ have distribution P_i, $i = 1, 2$. If $\mathscr{L}(\tau(Y_1)) = \mathscr{L}(\tau(Y_2))$, then $P_1 = P_2$.*

PROOF. Let Q be the common distribution of $\mathscr{L}(\tau(Y_i))$, $i = 1, 2$ and let μ be the invariant probability measure on G. Given a bounded measurable function f defined on $\mathbf{Y}$, the function

$$y \to \int_G f(gy)\mu(dg)$$

is invariant and is assumed to be measurable. Thus, this function can be written as $f_0(\tau(y))$ with f_0 defined on $\mathbf{X}$.

Now, using the invariance of P_i and the definition of Q, we have

$$\int f(y)P_i(dy) = \int f(gy)P_i(dy)$$
$$= \int\int f(gy)\mu(dg)P_i(dy) = \int f_0(\tau(y))P_i(dy) = \int f_0(x)Q(dx)$$

for $i = 1, 2$. Thus, for any bounded measurable function f,

$$\int f(y)P_1(dy) = \int f(y)P_2(dy)$$

so $P_1 = P_2$. □

Further properties of the matrix Δ and its distribution can be found in Eaton (1985).

7.3. Some asymptotics. Again consider Δ defined by (5.1) with p and q fixed, but n tending to infinity. The rows of X: $n \times q$, say $X_1', \ldots, X_n'$, are iid $N(0, I_q)$. Thus, by the strong law of large numbers,

$$n^{-1}\sum_1^n X_i X_i' \to \mathbf{E}X_1X_1' = I_q.$$

Therefore,

$$\sqrt{n}\,\Delta = \sqrt{n}\,Y(Y'Y + Z'Z)^{-1/2}$$
$$= \sqrt{n}\,Y\left(\sum_1^n X_i X_i'\right)^{-1/2} = Y\left(n^{-1}\sum_1^n X_i X_i'\right)^{-1/2}$$

converges almost surely to Y which is $N(0, I_p \otimes I_q)$. Thus we have:

PROPOSITION 7.5. *Let Γ be uniform on O_n and let Δ be any $p \times q$ subblock of the matrix Γ. Then $\sqrt{n}\,\Delta$ converges in distribution to a $N(0, I_p \otimes I_q)$ distribution as $n \to \infty$.*

We now turn to the question of the rate of convergence of the distribution of $\sqrt{n}\,\Delta$ to the normal. The result described here is from Diaconis, Eaton and Lauritzen (1987). Recall that for two probability measures P_1 and P_2, the variation distance between P_1 and P_2 is defined by

$$\|P_1 - P_2\| = 2\sup_B |P_1(B) - P_2(B)|,$$

where the sup ranges over the relevant σ-algebra. When P_1 and P_2 are both absolutely continuous with respect to a σ-finite measure λ, say $p_i = dP_i/d\lambda$, then

$$\|P_1 - P_2\| = \int |p_1 - p_2|\,d\lambda = 2\int\left(\frac{p_1}{p_2} - 1\right)^+ p_2\,d\lambda,$$

where $f^+ = f$ when f is positive and $f^+ = 0$ otherwise. Therefore,

$$\|P_1 - P_2\| \le 2 \sup_x \left(\frac{p_1(x)}{p_2(x)} - 1 \right)^+. \tag{7.6}$$

Here is one technical fact concerning variation distance which is used henceforth without comment. For P_1 and P_2 defined on a measurable space $(\mathbf{X}_1, \mathbf{B}_1)$, consider a measurable map f from $(\mathbf{X}_1, \mathbf{B}_1)$ to $(\mathbf{X}_2, \mathbf{B}_2)$. Then $fP_i = Q_i$ is defined on $(\mathbf{X}_2, \mathbf{B}_2)$ by

$$Q_i(B) = (fP_i)(B) = P_i(f^{-1}(B)), \qquad i = 1, 2.$$

Because f is measurable, $f^{-1}(\mathbf{B}_2) \subseteq \mathbf{B}_1$ so that

$$\|Q_1 - Q_2\| \le \|P_1 - P_2\|.$$

In other words, variation distance cannot be increased by measurable transformations.

It is inequality (7.6) which is used to bound the variation distance between the distribution of $\sqrt{n}\,\Delta$ and the $N(0, I_p \otimes I_q)$ distribution. We now proceed with some of the technical details. First note that the variation distance between $\mathscr{L}(\sqrt{n}\,\Delta)$ and $N(0, I_p \otimes I_q)$ is the same as the variation distance between $\mathscr{L}(\Delta)$ and $N(0, n^{-1}I_p \otimes I_q)$ because variation distance is invariant under one-to-one bimeasurable transformations. In the calculation below [from Diaconis and Freedman (1987)], we treat the case of $q = 1$ and $p = 2r$ as an even integer. Under this assumption, the density of Δ is [from (7.4)]

$$p_1(x) = \frac{(\sqrt{2\pi})^{-p} 2^{p/2} \Gamma(n/2)}{\Gamma((n-p)/2)} (1 - x'x)^{(n-p-2)/2} I_0(x'x)$$

for $x \in R^p$. Dividing $p_1(x)$ by the density of a $N(0, n^{-1}I_p)$ distribution, say $p_2(x)$, we have

$$\frac{p_1(x)}{p_2(x)} = \left(\frac{2}{n}\right)^{p/2} \frac{\Gamma(n/2)}{\Gamma((n-p)/2)} (1 - x'x)^{(n-p-2)/2} \exp\left[\frac{nx'x}{2}\right] I_0(x'x).$$

When $p \le n - 3$, this ratio is maximized for $x'x = (p+2)/n$ so the ratio is bounded above by

$$B = \left(\frac{2}{n}\right)^{p/2} \frac{\Gamma(n/2)}{\Gamma((n-p)/2)} \left(1 - \frac{p+2}{n}\right)^{(n-p-2)/2} \exp\left[\frac{p+2}{2}\right].$$

Since $p = 2r$ is even, it follows easily that

$$\log\left(\left(\frac{2}{n}\right)^{p/2} \frac{\Gamma(n/2)}{\Gamma((n-p)/2)} \right) = \sum_1^r \log\left(1 - \frac{2j}{n}\right)$$

$$= \sum_1^{r+1} \log\left(1 - \frac{2j}{n}\right) - \log\left(1 - \frac{p+2}{n}\right).$$

But

$$\sum_{1}^{r+1} \log\left(1 - \frac{2j}{n}\right) \le \int_{0}^{r+1} \log\left(1 - \frac{2x}{n}\right) dx$$

$$= -\frac{n-p-2}{2} \log\left(1 - \frac{p+2}{n}\right) - \frac{p+2}{2},$$

so

$$\log B \le -\log\left(1 - \frac{p+2}{n}\right).$$

Therefore,

$$B - 1 \le \frac{p+2}{n-p-2}.$$

Summarizing, we have:

PROPOSITION 7.6. [Diaconis and Freedman (1987)]. *If* $q = 1$ *and* $p \le n - 4$, *then the variation distance between* $\mathscr{L}(\sqrt{n}\,\Delta)$ *and a* $N(0, I_p)$ *distribution is bounded above by* $2(p+3)/(n-p-3)$.

PROOF. From inequality (7.6) when p is even, the variation distance is bounded above by $2(B-1)$ which is bounded above by $2(p+2)/(n-p-2) \le 2(p+3)/(n-p-3)$. The easy argument extending the bound to odd p is given in Diaconis and Freedman (1987). □

When $q > 1$, similar bounds have been established by Diaconis, Eaton and Lauritzen (1987), but the details are substantially more gory. Here is one version of the bound:

PROPOSITION 7.7. *For* $p + q \le n - 3$ *and* $t = \min\{p, q\}$, *the variation distance between* $\mathscr{L}(\sqrt{n}\,\Delta)$ *and a* $N(0, I_p \otimes I_q)$ *distribution is bounded above by*

$$\delta_n = 2\left[\exp\left[-c\log\left(1 - \frac{p+q+2}{n}\right)\right] - 1\right],$$

where

$$c = \frac{3t^2 + 5t}{8}.$$

Proof. See Diaconis, Eaton and Lauritzen (1987). □

It is possible to bound δ_n above by an expression of the form $a(p+q+2)/n$ (a is a constant) when $(p+q+2)/n$ is bounded away from 1. More explicitly, assume $(p+q+2)/n \leq \gamma < 1$ and set

$$\phi(x) = 2[\exp[-c\log(1-x)] - 1], \qquad 0 \leq x \leq \gamma,$$

where the constant c is given in Proposition 7.7. Because ϕ is increasing and convex on $[0, \gamma]$,

$$\phi(x) \leq \frac{\phi(\gamma)x}{\gamma}, \qquad 0 \leq x \leq \gamma.$$

Setting $a = a(\gamma) = \phi(\gamma)/\gamma$ yields the inequality

$$\delta_n \leq a(\gamma)\frac{p+q+2}{n}, \qquad \frac{p+q+2}{n} \leq \gamma.$$

This bound is qualitatively the same as that given in Proposition 7.6.

CHAPTER 8

Finite de Finetti Style Theorems

The purpose of this chapter is to introduce the ideas surrounding the so called finite de Finetti style theorems. Four examples, one of which comes from the classical de Finetti theorem and three related to the normal distribution, are discussed here. These examples are introduced by first describing the "infinite version" of a result and then moving to the "finite version." In all of these examples, the infinite version came first, followed by a finite version. However, recent work on finite versions has suggested new infinite versions; some of these are discussed in the next chapter.

8.1. The de Finetti theorem. We begin with a review of the classical de Finetti theorem for an exchangeable infinite sequence of 0-1 valued random variables. Let $\mathbf{X} = \{0, 1\}$ and for each integer n, $1 \leq n < +\infty$, let $\mathbf{X}^{(n)}$ be the n-fold product of $\mathbf{X}$ with itself. Given a probability P on the infinite product $\mathbf{X}^{\infty}$, $P^{(n)}$ denotes the projection of P onto $\mathbf{X}^{(n)}$. If $X = (X_1, X_2, \ldots)$ is a sequence of random variables with values in $\mathbf{X}^{\infty}$, then $X^{(n)}$ denotes the first n coordinates of X. Thus, if the probability law of X in $\mathbf{X}^{\infty}$ is P, written $\mathscr{L}(X) = P$, then

$$\mathscr{L}(X^{(n)}) = P^{(n)}.$$

Recall that P, a probability on $\mathbf{X}^{\infty}$, is called *exchangeable* if for each n, $P^{(n)}$ on $\mathbf{X}^{(n)}$ is exchangeable, that is, if $P^{(n)}$ is invariant under the action of the permutation group on $\mathbf{X}^{(n)}$. Equivalently, if $X \in \mathbf{X}^{\infty}$, then X is *exchangeable* if for each n, the random vector $X^{(n)}$ has a distribution which is invariant under permutations. As an example, let $Z = (Z_1, Z_2, \ldots)$ be a sequence of iid Bernoulli random variables with probability α of success and let P_α denote the distribution of Z on $\mathbf{X}^{\infty}$. Obviously P_α is exchangeable as is any mixture, over α, of P_α. That is, let μ be a probability measure defined on the Borel sets of $[0, 1]$ and define P_μ

on $\mathbf{X}^\infty$ by

$$P_\mu(B) = \int_0^1 P_\alpha(B)\mu(d\alpha) \tag{8.1}$$

for B in the σ-algebra of $\mathbf{X}^\infty$. Thus, for each n,

$$P_\mu^{(n)}(B) = \int_0^1 P_\alpha^{(n)}(B)\mu(d\alpha) \tag{8.2}$$

for relevant sets B. These two equations are often written as

$$P_\mu = \int_0^1 P_\alpha \mu(d\alpha) \tag{8.3}$$

and

$$P_\mu^{(n)} = \int P_\alpha^{(n)} \mu(d\alpha), \tag{8.4}$$

a notation which is adopted here. Thus, P_μ given in (8.3) is exchangeable. The important observation of de Finetti (1931) is:

THEOREM 8.1. *Suppose P on $\mathbf{X}^\infty$ is exchangeable. Then there is a unique probability measure μ on $[0,1]$ such that*

$$P = \int_0^1 P_\alpha \mu(d\alpha). \tag{8.5}$$

One consequence of (8.5) is that for each positive integer k,

$$P^{(k)} = \int_0^1 P_\alpha^{(k)} \mu(d\alpha). \tag{8.6}$$

In other words, all of the marginal distributions of P have the representation (8.6). Now, *fix* a finite integer n and assume $P^{(n)}$ on $\mathbf{X}^{(n)}$ is exchangeable. Thus all of the lower dimensional marginals, say $P^{(k)}$ with $1 \le k < n$, are exchangeable. It seems natural to ask if the $P^{(k)}$ have the representation (8.6). The answer is no; an example is given below. However, what is true is that the $P^{(k)}$ "almost" have such a representation when n is a lot bigger than k. The problem is to make this precise. We now turn to a careful discussion of this problem which was solved by Diaconis and Freedman (1980).

The sample space $\mathbf{X}^{(n)}$ consists of n dimensional vectors, which we write as column vectors

$$x = \begin{pmatrix} x_1 \\ x_2 \\ \vdots \\ x_n \end{pmatrix},$$

where each x_i is 0 or 1. The group of $n \times n$ permutation matrices $\mathscr{P}_n$ acts on the left of $\mathbf{X}^{(n)}$. Consider a probability measure $P^{(n)}$ on $\mathbf{X}^{(n)}$ which is exchangeable, that is, $P^{(n)}$ satisfies

(8.7) $$gP^{(n)} = P^{(n)}, \qquad g \in \mathscr{P}_n,$$

or equivalently,

$$\mathscr{L}(X^{(n)}) = \mathscr{L}(gX^{(n)}), \qquad g \in \mathscr{P}_n,$$

where $P^{(n)} = \mathscr{L}(X^{(n)})$. The results of Example 4.2 give us a representation for $P^{(n)}$. A cross section in this example is

$$\mathbf{Y} = \{y_0, y_1, \ldots, y_n\},$$

where y_i has its first i elements equal to 1 and the remaining elements are 0. Thus $X^{(n)}$ has a representation as $X^{(n)} = UY$ where U is uniform on $\mathscr{P}_n$, Y is independent of U and Y has an arbitrary distribution on $\mathbf{Y}$. Let H_i denote the distribution of Uy_i on $\mathbf{X}^{(n)}$. Obviously H_i is the uniform distribution on the orbit

$$\{gy_i | g \in \mathscr{P}_n\}$$

and H_i puts mass $\binom{n}{i}^{-1}$ on each point in this orbit. Let

$$p_i = \text{Prob}\{Y = y_i\}.$$

From the representation $X^{(n)} = UY$, it is clear that

(8.8) $$P^{(n)} = \sum_{i=0}^{n} p_i H_i.$$

Conversely, any probability measure of the form

$$\sum_{i=0}^{n} p_i H_i, \qquad 0 \leq p_i, \Sigma p_i = 1,$$

is exchangeable. Further, the representation is unique because the H_i are mutually singular. Summarizing we have:

THEOREM 8.2. *In order that $P^{(n)}$ on $\mathbf{X}^{(n)}$ be exchangeable it is necessary and sufficient that*

(8.9) $$P^{(n)} = \sum_{i=0}^{n} p_i H_i$$

for some $p_i \geq 0$, $\Sigma p_i = 1$. The representation is unique.

It is clear that the set of exchangeable probabilities on $\mathbf{X}^{(n)}$ is a convex set. Theorem 8.2 shows that the extreme points of this convex set are $H_0, H_1, \ldots, H_n$. Now, focus on the exchangeable probability H_1 and let $\mathscr{L}(X^{(n)}) = H_1$. Consider the possibility of representing H_1 in the form (8.6), that is, suppose

(8.10) $$H_1 = \int_0^1 P_\alpha^{(n)} \mu(d\alpha)$$

for some μ where $P_\alpha^{(n)}$ is the probability measure for iid Bernoullis with success probability α. The claim is that (8.10) cannot hold for any μ. On the contrary, if

(8.10) holds, observe that

$$1/n = \mathbf{E}X_1^{(n)} = \int_0^1 \alpha\mu(d\alpha)$$

and

$$0 = \mathbf{E}X_1^{(n)}X_2^{(n)} = \int_0^1 \alpha^2\mu(d\alpha).$$

The second equation implies that $\mu(\{0\}) = 1$ and this contradicts the first equation. This shows Theorem 8.1 is false for every finite n.

Again assume $P^{(n)} = \mathscr{L}(X^{(n)})$ is an exchangeable probability on $\mathbf{X}^{(n)}$. As usual, $X^{(k)}$ is the vector of the first k coordinates of $X^{(n)}$ where $P^{(k)} = \mathscr{L}(X^{(k)})$. Obviously $P^{(k)}$ is an exchangeable probability on $\mathbf{X}^{(k)}$ and $P^{(k)}$ is the "projection" of $P^{(n)}$ down to $\mathbf{X}^{(k)}$. More precisely, let π be the $k \times n$ matrix defined by

$$\pi = (I_k \quad 0)\colon k \times n,$$

where I_k is the $k \times k$ identity matrix. Obviously

$$\pi X^{(n)} = X^{(k)}$$

so

$$\pi P^{(n)} = P^{(k)},$$

where

$$(\pi P^{(n)})(B) = P^{(n)}(\pi^{-1}(B))$$

for subsets B of $\mathbf{X}^{(k)}$. A main result in Diaconis and Freedman (1980) shows that

$$\Delta_{k,n} = \inf_{\mu}\left\| P^{(k)} - \int_0^1 P_\alpha^{(k)}\mu(d\alpha)\right\| \le 4k/n, \tag{8.11}$$

where $\|\cdot\|$ denotes variation distance (as discussed in Chapter 7) and the inf is over all the Borel measures on $[0,1]$. The interpretation of (8.11) is that when $P^{(k)}$ is the projection of an exchangeable probability on $\mathbf{X}^{(n)}$, then $P^{(k)}$ is within $4k/n$ of some mixture of iid Bernoullis. The basic step in the proof of (8.11) is the following:

THEOREM 8.3. *The variation distance between πH_i and $P_\alpha^{(k)}$ with $\alpha = i/n$ is bounded above by $4k/n$.*

PROOF. With $\mathscr{L}(X^{(k)}) = \pi H_i$, $X^{(k)}$ is the outcome of k draws made without replacement from an urn with i 1's and $n - i$ 0's. But $P_\alpha^{(k)}$ represents the probability measure of k draws made with replacement from the same urn. Bounding the variation distance between πH_i and $P_\alpha^{(k)}$, which involves some calculus, is carried out in Lemma 6 of Diaconis and Freedman (1980). □

THEOREM 8.4. *Given an exchangeable $P^{(n)}$ on $\mathbf{X}^{(n)}$ and $P^{(k)} = \pi P^{(n)}$, Equation* (8.11) *holds.*

PROOF. First use Theorem 8.2 to write

$$P^{(n)} = \sum_{i=0}^{n} p_i H_i$$

so that

$$P^{(k)} = \pi P^{(n)} = \sum_{i=0}^{n} p_i \pi H_i.$$

Let μ_0 be the probability on $[0,1]$ which puts mass p_i at the point $\alpha_i = i/n$, $i = 0, \ldots, n$. Then

$$\Delta_{k,n} = \inf_{\mu} \left\| P^{(k)} - \int_0^1 P_\alpha^{(k)} \mu(d\alpha) \right\| \leq \left\| P^{(k)} - \int_0^1 P_\alpha^{(k)} \mu_0(d\alpha) \right\|$$

$$= \left\| \sum_0^n p_i \pi H_i - \sum_0^n p_i P_{\alpha_i}^{(k)} \right\| \leq \sum_0^n p_i \left\| \pi H_i - P_{\alpha_i}^{(k)} \right\| \leq 4k/n,$$

where the last inequality follows from Theorem 8.3. Then next to the last inequality is a consequence of the fact that variation distance is a norm and hence is a convex function. □

The argument given above shows that to bound $\Delta_{k,n}$ in (8.11), it is sufficient (and necessary) to bound $\Delta_{k,n}$ when $P^{(k)}$ is one of the projected extreme points πH_i, $i = 0, \ldots, n$. This type of argument is used in all of the examples in this and the next chapter. Theorem 8.4 is often called a finite style de Finetti theorem because n and k are both fixed and finite. This result can be used to provide an easy proof of the infinite de Finetti theorem. For example, see Theorem 14 in Diaconis and Freedman (1980) where the sort of argument used above provides an easy proof of the Hewitt–Savage (1955) generalization of the de Finetti theorem. An interesting related paper is Dubins and Freedman (1979).

Finally, a few remarks about extendability. Theorem 8.4 concerns those $P^{(k)}$ on $\mathbf{X}^{(k)}$ which are *n-extendable* in the sense that there exists an exchangeable $P^{(n)}$ on $\mathbf{X}^{(n)}$ such that

$$P^{(k)} = \pi P^{(n)}.$$

Thus, an n-extendability assumption on $P^{(k)}$ is equivalent to saying that $P^{(k)}$ is the projection of some exchangeable $P^{(n)}$ on $\mathbf{X}^{(n)}$. This latter condition is a bit more convenient and will appear throughout this and the next chapter. However, the reader should keep the equivalence in mind since n-extendability sometimes is a bit easier to think about.

The results of this section show that if $P^{(k)}$ is n-extendable for all large n, then $P^{(k)}$ has the representation (8.6). However, if $P^{(k)}$ is n-extendable for some fixed n, then (8.6) need not hold, but when n is much bigger than k, then (8.6) almost holds in the sense of Theorem 8.4.

8.2. Orthogonally invariant random vectors. The material in this section is related to Example 4.3. For $1 \leq n \leq \infty$, let R^n denote n dimensional coordinate space. Given $X = (X_1, X_2, \ldots)$ in R^∞, X has an *orthogonally invariant* distribution if for each finite n,

$$X^{(n)} = \begin{pmatrix} X_1 \\ X_2 \\ \vdots \\ X_n \end{pmatrix}$$

has an O_n invariant distribution. If $X \in R^\infty$ and $P = \mathscr{L}(X)$, we say P is *orthogonally invariant* if X is orthogonally invariant. Of course, this means that for each finite n, the projected measures

$$P^{(n)} = \mathscr{L}(X^{(n)})$$

are O_n invariant.

For example, if $Z = (Z_1, Z_2, \dots)$ has iid coordinates which are $N(0, \sigma^2)$, let P_σ denote the probability on R^∞ of Z. Then $P_\sigma^{(n)}$ is the joint distribution of

$$Z^{(n)} = \begin{pmatrix} Z_1 \\ Z_2 \\ \vdots \\ Z_n \end{pmatrix}.$$

In other words,

$$\mathscr{L}(Z^{(n)}) = N(0, \sigma^2 I_n)$$

for $0 \le \sigma < +\infty$, so P_σ is orthogonally invariant. Given any probability μ on $[0, \infty)$, it is clear that

$$P_\mu = \int_0^\infty P_\sigma \mu(d\sigma) \tag{8.12}$$

is orthogonally invariant. Probabilities of the form (8.12) are called scaled mixtures of normals. When (8.12) holds, then for each n,

$$P_\mu^{(n)} = \int_0^\infty P_\sigma^{(n)} \mu(d\sigma). \tag{8.13}$$

In the present setting, here is the "infinite theorem."

THEOREM 8.5. *P on R^∞ is orthogonally invariant iff P has the representation* (8.12). *Further the representation in* (8.12) *is unique.*

This result is commonly attributed to Schoenberg, but see Section 6 in Diaconis and Freedman (1987). A proof of this theorem, based on the "finite version" given below, can be found in Theorem 3 in Diaconis and Freedman (1987). The representation has been rediscovered in a number of different contexts, for example, see Hill (1969), Andrews and Mallows (1974) and Eaton (1981). The uniqueness part of the theorem follows easily from the uniqueness of Laplace transforms because (8.12) implies

$$P_\mu^{(1)} = \int_0^\infty P_\sigma^{(1)} \mu(d\sigma).$$

Thus $P_{\mu}^{(1)}$ has characteristic function

$$t \to \int_0^{\infty} \exp\left[-\tfrac{1}{2}\sigma^2 t^2\right] \mu(d\sigma).$$

Therefore, if μ_1 and μ_2 both represent P_1, they have the same Laplace transforms and hence are equal.

We now turn to a finite version of Theorem 8.5. Fix a positive integer n and let $P^{(n)}$ be an O_n-invariant probability on R^n. Given $r \geq 0$, let H_r denote the uniform distribution on

$$\{x | x \in R^n, \|x\| = r\},$$

the sphere of radius r in R^n. Naturally H_0 is the probability degenerate at $0 \in R^n$. Clearly each H_r is O_n-invariant. The arguments given in Chapter 4 establish:

THEOREM 8.6. *A probability $P^{(n)}$ on $R^{(n)}$ is O_n-invariant iff for some Borel measure μ on $[0, \infty)$,*

$$P^{(n)} = \int_0^{\infty} H_r \mu(dr). \tag{8.14}$$

It is clear that the O_n-invariant probability H_1 cannot be represented in the form (8.12). Thus, Theorem 8.5 is false for any finite integer n. To establish an analog of Theorem 8.4 in the present context, fix an integer $k < n$ and let $P^{(k)}$ be the probability measure of the first k coordinates of $X^{(n)}$ where $P^{(n)} = \mathscr{L}(X^{(n)})$. Further, let

$$\pi = (I_k \quad 0) \colon k \times n$$

be a $k \times n$ real matrix so

$$X^{(k)} = \pi X^{(n)}$$

and

$$P^{(k)} = \pi P^{(n)}.$$

The main result below, due to Diaconis and Freedman (1987), shows that $P^{(k)}$ is close to a scale mixture of normals in the following sense:

THEOREM 8.7. *Assume $P^{(n)}$ is O_n-invariant and $k \leq n - 4$. Then, with $P^{(k)} = \pi P^{(n)}$,*

$$\Delta_{k,n} = \inf_{\mu} \left\| P^{(k)} - \int_0^{\infty} P_{\sigma}^{(k)} \mu(d\sigma) \right\| \leq \frac{2(k+3)}{n-k-3}, \tag{8.15}$$

where $\|\cdot\|$ denotes variation distance and the inf *is over all Borel measures on $[0, \infty)$.*

The proof of this theorem follows much the same lines as the proof of Theorem 8.4. Equation (8.15) is first established for πH_r and then (8.14) is used for the general case.

Theorem 8.8. *Inequality* (8.15) *holds for* $P^{(k)} = \pi H_r$ *for each* $r \geq 0$.

Proof. For $r = 0$, the result is obvious. For $r > 0$, H_r is the probability measure of the random vector

$$X^{(n)} = rU^{(n)},$$

where $U^{(n)}$ is uniform on the sphere of radius 1 in R^n. Thus, πH_r is the distribution of

$$\pi X^{(n)} = rU^{(k)}.$$

Taking $p = k$ in Proposition 7.6 shows that

$$-\left\| \mathscr{L}(\sqrt{n}\, U^{(k)}) - N(0, I_k) \right\| \leq 2(k+3)/(n-k-3).$$

Because variation distance is invariant under one-to-one bimeasurable transformations, this implies that

$$\left\| \mathscr{L}(rU^{(k)}) - N\left(0, n^{-1}r^2 I_k\right) \right\| \leq 2(k+3)/(n-k-3). \tag{8.16}$$

Hence (8.15) holds for πH_r because $\pi H_r = \mathscr{L}(rU^{(k)})$. □

Proof of Theorem 8.7. Because $P^{(n)}$ is O_n-invariant, (8.14) implies that

$$P^{(k)} = \pi P^{(n)} = \int_0^\infty \pi H_r \mu_0(dr)$$

for some μ_0. Thus, using (8.16),

$$\begin{aligned}
\inf_\mu & \left\| P^{(k)} - \int_0^\infty N\left(0, r^2 I_k\right) \mu(dr) \right\| \\
&\leq \left\| P^{(k)} - \int_0^\infty N\left(0, n^{-1}r^2 I_k\right) \mu_0(dr) \right\| \\
&= \left\| \int_0^\infty \pi H_r \mu_0(dr) - \int_0^\infty N\left(0, n^{-1}r^2 I_k\right) \mu_0(dr) \right\| \\
&\leq \int_0^\infty \left\| \pi H_r - N\left(0, n^{-1}r^2 I_k\right) \right\| \mu_0(dr) \leq 2(k+3)/(n-k-3). \qquad \square
\end{aligned}$$

The essentials of the argument are much the same as they were in Section 8.1, namely, the set of O_n-invariant probabilities is a convex set with extreme points H_r, $r \geq 0$. Thus, to approximate $P^{(k)}$ well by a scale mixture of normals, it is sufficient to approximate πH_r well (in this case, uniformly) by scaled normals. This is what Theorem 8.8 together with Proposition 7.6 does.

The remarks concerning extendability made at the end of the previous section apply here. In particular, if $P^{(k)}$ on R^k is O_k-invariant and if $P^{(k)}$ is n-extendable (that is, $P^{(k)} = \pi P^{(n)}$ for some O_n-invariant $P^{(n)}$ on R^n), then $P^{(k)}$ is within $2(k+3)/(n-k-3)$ of some scale mixture of normals. This is just a restatement of Theorem 8.7.

8.3. Orthogonally invariant random matrices. Here, the results of the previous section are extended to the matrix case. First a bit of notation is needed. Fix a positive integer q and let $\mathscr{L}_{q,n}$ be the vector space of all real $n \times q$ matrices, $1 \le n \le +\infty$. Given a random matrix X in $\mathscr{L}_{q,\infty}$, let $X^{(n)}$: $n \times q$ for $1 \le n < +\infty$ denote the matrix in $\mathscr{L}_{q,n}$ consisting of the first n rows of X. If $P = \mathscr{L}(X)$ is the distribution of X, then $P^{(n)}$ denotes the distribution of $X^{(n)}$. The group O_n acts on $\mathscr{L}_{q,n}$ via matrix multiplication on the left:

$$x \to gx, \qquad x \in \mathscr{L}_{q,n}, \qquad g \in O_n.$$

A probability P on $\mathscr{L}_{q,\infty}$ is *left-orthogonally invariant* if for each finite n,

$$P^{(n)} = gP^{(n)}, \qquad g \in O_n.$$

Thus, if $\mathscr{L}(X) = P$ and P is left-orthogonally invariant, then

$$\mathscr{L}\big(gX^{(n)}\big) = \mathscr{L}\big(X^{(n)}\big), \qquad g \in O_n,$$

for each finite n.

As an example, consider Z in $\mathscr{L}_{q,\infty}$ whose rows $Z_1', Z_2', \dots$ are iid $N_q(0, \alpha^2)$ where α is a $q \times q$ positive semidefinite matrix. Then

$$\mathscr{L}\big(Z^{(n)}\big) = N\big(0, I_n \otimes \alpha^2\big)$$

with $\otimes$ denoting the Kronecker product. Let $P_\alpha = \mathscr{L}(Z)$ so P_α is obviously left-orthogonally invariant. Further, given any probability measure μ on the set $\mathbf{S}$ of $q \times q$ positive semidefinite matrices, the probability

$$P_\mu = \int P_\alpha \mu(d\alpha) \tag{8.17}$$

is also left-orthogonally invariant since

$$P_\mu^{(n)} = \int P_\alpha^{(n)} \mu(d\alpha). \tag{8.18}$$

The converse of this observation, established in Dawid (1977), is:

THEOREM 8.9. *Assume P on $\mathscr{L}_{q,\infty}$ is left-orthogonally invariant. Then P has the representation* (8.17). *Further, the representation is unique.*

This "infinite" theorem is usually stated as "P is left-orthogonally invariant iff P is a covariance mixture of normals." The uniqueness of μ is proved in the same way it is proved in the case $q = 1$. Theorem 8.9 can be proved using the finite version of this theorem to which we now turn.

As in the two previous sections, now fix a finite n and consider $P^{(n)}$ on $\mathscr{L}_{q,n}$ which is left-orthogonally invariant. Our first task is to apply Theorem 4.1 to the case at hand. The group O_n acts on $\mathscr{L}_{q,n}$. To specify a cross section in $\mathscr{L}_{q,n}$, let

$$\mathbf{Y} = \left\{ x \middle| x \in \mathscr{L}_{q,n}, x = \begin{pmatrix} \alpha \\ 0 \end{pmatrix}, \alpha \in \mathbf{S} \right\},$$

and define τ on $\mathscr{L}_{q,n}$ to $\mathbf{Y}$ by

$$\tau(x) = \begin{pmatrix} (x'x)^{1/2} \\ 0 \end{pmatrix}.$$

Here, $(x'x)^{1/2}$ denotes the unique positive semidefinite square root of $x'x \in \mathbf{S}$. That $\mathbf{Y}$ is a measurable cross section (according to Definition 4.1) is easily checked. Theorem 4.3 yields:

THEOREM 8.10. *For $\alpha \in \mathbf{S}$, let H_α denote the distribution of*

$$U\begin{pmatrix}\alpha\\0\end{pmatrix},$$

where U is uniform on O_n and $\begin{pmatrix}\alpha\\0\end{pmatrix}$ is in $\mathbf{Y}$. Then $P^{(n)}$ on $\mathscr{L}_{q,n}$ is left-orthogonally invariant iff

$$P^{(n)} = \int H_\alpha \mu(d\alpha) \tag{8.19}$$

for some probability μ on $\mathbf{S}$.

PROOF. Apply Theorem 4.3 with $H_\alpha = \mu_y$ and $\mu = Q$. □

Now, let π denote the $k \times n$ matrix

$$\pi = (I_k \quad 0),$$

where $k < n$. If $P^{(n)} = \mathscr{L}(X^{(n)})$, then

$$P^{(k)} = \pi P^{(n)} = \mathscr{L}(\pi X^{(n)}).$$

To establish a finite theorem for $P^{(k)}$, we first establish a finite theorem for πH_α and then use (8.19). □

THEOREM 8.11. *For $k + q \le n - 3$, the variation distance between πH_α and the normal distribution $N(0, n^{-1}I_k \otimes \alpha^2)$ is bounded above by δ_n given in Proposition 7.7.*

PROOF. Recall that H_α is the distribution of

$$U\begin{pmatrix}\alpha\\0\end{pmatrix} = U\begin{bmatrix}I_q\\0\end{bmatrix}\alpha,$$

where U is uniform on O_n. Thus, πH_α is the distribution of

$$\pi U\begin{pmatrix}\alpha\\0\end{pmatrix} = (I_k \quad 0)U\begin{bmatrix}I_q\\0\end{bmatrix}\alpha = \Delta\alpha,$$

where Δ is the $k \times q$ upper left corner of U. Proposition 7.7 implies that

$$\left\|\mathscr{L}(\Delta) - N\left(0, n^{-1}I_k \otimes I_q\right)\right\| \le \delta_n.$$

Since $\pi H_\alpha = \mathscr{L}(\Delta\alpha)$, the result follows. □

The following finite theorem is from Diaconis, Eaton and Lauritzen (1987).

THEOREM 8.12. *Suppose $P^{(n)}$ on $\mathscr{L}_{q,n}$ is left-orthogonally invariant. If $k + q \le n - 3$, then*

$$\inf_\mu \left\| P^{(k)} - \int N\left(0, I_k \otimes \alpha^2\right)\mu(d\alpha)\right\| \le \delta_n, \tag{8.20}$$

where the inf *ranges over all probabilities on* **S** *and* δ_n *is given in Proposition* 7.7 (*with p replaced by k*).

PROOF. Since $P^{(n)}$ is left-orthogonally invariant, we can write

$$P^{(n)} = \int H_\alpha \mu_0(d\alpha)$$

for some probability μ_0 on **S**. Therefore,

$$P^{(k)} = \pi P^{(n)} = \int \pi H_\alpha \mu_0(d\alpha).$$

Since variation distance is a convex function, Theorem 8.11 yields

$$\left\| P^{(k)} - \int N(0, n^{-1}I_k \otimes \alpha^2)\mu_0(d\alpha) \right\|$$

$$= \left\| \int \left[\pi H_\alpha - N(0, n^{-1}I_k \otimes \alpha^2)\right] \mu_0(d\alpha) \right\|$$

$$\leq \int \left\| \pi H_\alpha - N(0, n^{-1}I_k i \otimes \alpha^2) \right\| \mu_0(d\alpha) \leq \delta_n.$$

Hence (8.20) holds. □

The comments concerning extendability made at the end of the last section are valid here. Of course, extendability refers to increasing n with fixed k and q.

8.4. A linear model example. Some new considerations arise when we try to formulate a finite version of an "infinite" theorem described in Smith (1981). To describe the infinite result, let R^n, $1 \leq n \leq \infty$, denote n dimensional coordinate space and for each finite n, let

$$O_n(e) = \{g | g \in O_n, ge = e\},$$

where e is the vector of 1's in R^n. Let $Z = (Z_1, Z_2, \ldots) \in R^\infty$ have coordinates which are iid $N(m, \sigma^2)$ where $m \in R^1$ and $\sigma \geq 0$. Thus the distribution of

$$Z^{(n)} = \begin{pmatrix} Z_1 \\ Z_2 \\ \vdots \\ Z_n \end{pmatrix}$$

is $N(me, \sigma^2 I_n)$. Clearly

$$\mathscr{L}(Z^{(n)}) = \mathscr{L}(gZ^{(n)}), \qquad g \in O_n(e).$$

$P_{m,\sigma}$ denotes the distribution of Z on R^∞ and

$$\mathscr{L}(Z^{(n)}) = P_{m,\sigma}^{(n)} = N(me, \sigma^2 I_n).$$

Given a probability μ on $R^1 \times [0, \infty)$, let

$$P_\mu = \int\int P_{m,\sigma} \mu(dm, d\sigma) \tag{8.21}$$

so P_μ is a translation-scale mixture of iid normals. Thus the projection of P_μ on R^n is given by

$$P_\mu^{(n)} = \int\int P_{m,\sigma}^{(n)} \mu(dm, d\sigma).$$

Clearly $gP_\mu^{(n)} = P_\mu^{(n)}$ for $g \in O_n(e)$.

THEOREM 8.13 [Smith (1981)]. *Let P be any probability on R^∞. Then the projection of P on R^n, say $P^{(n)}$, is $O_{(n)}(e)$-invariant for all $n = 1, 2, \ldots$ iff P has the representation* (8.21).

To describe a finite version of this result, we first need a representation for $P^{(n)}$ defined on R^n (fixed n) which is $O_n(e)$-invariant. Here is a convenient cross section for this example. Fix the vector

$$x_0 = \frac{1}{\sqrt{2}} \begin{pmatrix} 1 \\ -1 \\ 0 \\ \vdots \\ 0 \end{pmatrix}$$

and let

$$\mathbf{Y} = \{x | x \in R^n, \, x = \sigma x_0 + me; \, \sigma \geq 0, \, m \in R^1\}.$$

Define τ on R^n to $\mathbf{Y}$ by

$$\tau(x) = \|x - \bar{x}e\| x_0 + \bar{x}e,$$

where

$$\bar{x} = n^{-1} \sum_1^n x_i$$

and $\|\cdot\|$ denotes standard Euclidean distance. That $\mathbf{Y}$ is a measurable cross section is easily verified. Given $\sigma \geq 0$ and $m \in R^1$, let $H_{m,\sigma}$ be the distribution of

$$\sigma U x_0 + me,$$

where U is uniform on $O_n(e)$. Note that the random vector Ux_0 has a uniform distribution on

$$\{x | x \in R^n, \, \|x\| = 1, \, x'e = 0\}.$$

THEOREM 8.14. *Let $P^{(n)}$ be a probability on R^n. Then, $P^{(n)}$ is $O_n(e)$-invariant iff*

$$P^{(n)} = \int\int H_{m,\sigma} \mu(dm, d\sigma)$$

for some probability μ on $R^1 \times [0, \infty)$.

PROOF. This is an easy application of Theorem 4.3. □

As usual, we use

$$\pi = (I_k \quad 0) : k \times n$$

to project down from R^n to R^k with $k < n$. The next step in the argument is to approximate $\pi H_{m,\sigma}$ by some normal distribution. The approximation is based on the following:

Lemma 8.1. *For U uniform on $O_n(e)$, $\pi U x_0$ is distributed as AV where:*

(i) *V is distributed as the first k coordinates of a random vector which has a uniform distribution on*

$$\{x | x \in R^{n-1}, \|x\| = 1\}.$$

(ii) *The $k \times k$ fixed matrix A is given by*

$$A = (\pi Q_0 \pi')^{1/2}$$

with

$$Q_0 = I_n - n^{-1} e e'.$$

Proof. See Proposition A.1 in Diaconis, Eaton and Lauritzen (1987). □

Theorem 8.15. *For $k \le n - 5$,*

$$\left\| \pi H_{m,\sigma} - N\left(m\pi e, (n-1)^{-1}\sigma^2 I_k\right) \right\| \le \beta_n, \tag{8.22}$$

where

$$\beta_n = 2\frac{k+3}{n-k-4} + 2\left[(\det A)^{-1} - 1\right]. \tag{8.23}$$

Proof. The probability $\pi H_{m,\sigma}$ is the law of

$$\sigma \pi U x_0 + m\pi e,$$

which, according to Lemma 8.1, is the same as the law of

$$\sigma A V + m\pi e.$$

Here, A and V are as defined in Lemma 8.1. For notational convenience, let W be $N(0, I_k)$. Thus, the left side of (8.22) is

$$\begin{aligned}
&\left\| \mathscr{L}(\sigma A V + m\pi e) - \mathscr{L}\left((n-1)^{-1/2}\sigma W + m\pi e\right) \right\| \\
&\quad \le \left\| \mathscr{L}(AV) - \mathscr{L}\left((n-1)^{-1/2} W\right) \right\| \\
&\quad \le \left\| \mathscr{L}(AV) - \mathscr{L}\left((n-1)^{-1/2} A W\right) \right\| \\
&\qquad + \left\| \mathscr{L}\left((n-1)^{-1/2} A W\right) - \mathscr{L}\left((n-1)^{-1/2} W\right) \right\| \\
&\quad \le \left\| \mathscr{L}(V) - \mathscr{L}\left((n-1)^{-1/2} W\right) \right\| + \left\| \mathscr{L}(AW) - \mathscr{L}(W) \right\|.
\end{aligned}$$

But, Proposition 7.6 (with n replaced by $n - 1$) yields

$$\left\| \mathscr{L}(V) - \mathscr{L}\left((n-1)^{-1/2}W\right) \right\| \leq 2\frac{k+3}{n-k-4}$$

for $k \leq n - 5$. Because all the eigenvalues of A are less than or equal to 1, the easily established inequality

$$\| \mathscr{L}(AW) - \mathscr{L}(W) \| \leq 2\left[(\det A)^{-1} - 1\right]$$

completes the proof. □

Finally, we come to the finite version of Theorem 8.13.

Theorem 8.16. *Given $P^{(n)}$ on R^n which is $O_n(e)$-invariant and $k \leq n - 5$, let $P^{(k)} = \pi P^{(n)}$. Then*

$$\inf_{\mu} \left\| P^{(k)} - \int\int N\left(m\pi e, \sigma^2 I_k\right)\mu(dm, d\sigma) \right\| \leq \beta_n, \tag{8.24}$$

where the inf *ranges over all probabilities on $R^1 \times [0, \infty)$ and β_n is given in* (8.23).

Proof. Since $P^{(n)}$ is $O_n(e)$-invariant, Theorem 8.14 yields

$$P^{(k)} = \pi P^{(n)} = \int\int \pi H_{m,\sigma}\mu_0(dm, d\sigma)$$

for some μ_0. Thus

$$\begin{aligned}
&\inf_{\mu} \left\| P^{(k)} - \int\int N\left(m\pi e, \sigma^2 I_k\right)\mu(dm, d\sigma) \right\| \\
&\quad \leq \left\| \int\int \left[\pi H_{m,\sigma} - N\left(m\pi e, (n-1)^{-1}\sigma^2 I_k\right)\right] \mu_0(dm, d\sigma) \right\| \\
&\quad \leq \int\int \left\| \pi H_{m,\sigma} - N\left(m\pi e, (n-1)^{-1}\sigma^2 I_k\right) \right\| \mu_0(dm, d\sigma) \leq \beta_n.
\end{aligned}$$

The final inequality follows from Theorem 8.15. □

The upper bound β_n in (8.23) and (8.24) consists of two parts. The argument used to prove Theorem 8.15 pinpoints the origin of the two pieces. The first piece is from a routine application of Proposition 7.6 which we understand fairly well. The second piece arises because the group in question leaves the subspace span $\{e\}$ fixed so that previous arguments must be modified by dropping down one dimension. The reduction in dimension introduces the $k \times k$ matrix A which appears in the bound as

$$2\left[(\det A)^{-1} - 1\right].$$

A routine calculation shows that

$$\det A^2 = 1 - \frac{k}{n}$$

so

$$(\det A)^{-1} - 1 \le \left(1 - \frac{k}{n}\right)^{-1/2} - 1 \le \frac{k}{n-k}$$

for $k \le n - 5$. Thus β_n is bounded above by

$$4\frac{k+3}{n-k-4}$$

for $k \le n - 5$. This bound is of the same type as obtained for the previous finite theorems (a constant times k/n for k/n bounded away from 1). From this, we conclude that the situation considered in this section is qualitatively the same as the situation in Section 8.2.

The finite result of this section is from Diaconis, Eaton and Lauritzen (1987) where a multivariate version of Theorem 8.16 is also proved. The previous remarks on extensions are of course valid for the situation of this section.

CHAPTER 9

Finite de Finetti Style Theorems for Linear Models

The implications of invariance assumptions for linear models are investigated here via a finite de Finetti style theorem. Before discussing linear models, we describe a general method for approximating projected measures. Of course, the origins of the method are in the four examples described in Chapter 8. All of the material in this chapter is from Diaconis, Eaton and Lauritzen (1987).

9.1. Approximating extendable probabilities. Consider a measurable space $(\mathbf{X}_2, \mathbf{B}_2)$ which is acted on by a compact group G. Let $\mathscr{P}$ be the set of all G-invariant probability measures defined on $\mathbf{B}_2$. The symbol U denotes a random element of G which has the uniform distribution on G. For each $x \in \mathbf{X}_2$, let

$$H_x = \mathscr{L}(Ux).$$

It was pointed out in Chapter 4 that each $P \in \mathscr{P}$ has the representation

$$P = \int H_x P(dx). \tag{9.1}$$

In other words, every element of the convex set $\mathscr{P}$ can be represented as an average of the H_x's. It is clear that for any $x \in \mathbf{X}_2$,

$$kH_{gx} = H_x \quad \text{for } k, g \in G,$$

because

$$\mathscr{L}(U) = \mathscr{L}(k^{-1}Ug) \quad \text{for } k, g \in G.$$

Now, let Q_x be an "approximation" to H_x. In applications, Q_x is often taken to be a normal distribution whose mean and covariance match those of H_x. Further, in all the applications that I know Q_x satisfies

$$kQ_{gx} = Q_x \quad \text{for } k, g \in G,$$

just as H_x does.

Next consider a second measurable space $(\mathbf{X}_1, \mathbf{B}_1)$ and a measurable map

$$\pi\colon \mathbf{X}_2 \to \mathbf{X}_1.$$

Think of a map π as a projection (as it was in Chapter 8). Then, let

$$\mathscr{P}_{21} = \{\pi P | P \in \mathscr{P}\}$$

be the set of projected invariant measures on $(\mathbf{X}_1, \mathbf{B}_1)$. Because of (9.1),

$$\pi P = \int \pi H_x P(dx) \tag{9.2}$$

so that all the elements in the convex set $\mathscr{P}_{21}$ are averages of the πH_x's. The elements of $\mathscr{P}_{21}$ are just those probabilities P_1 on $(\mathbf{X}_1, \mathbf{B}_1)$ which have G invariant extensions to $(\mathbf{X}_2, \mathbf{B}_2)$, extensions in the sense that there is a $P \in \mathscr{P}$ such that $\pi P = P_1$.

The following result, which captures the essence of the argument used throughout Chapter 8, provides an upper bound on the variation distance between (i) an element πP of $\mathscr{P}_{21}$ and (ii) the closest approximation to πP based on averages of the πQ_x's.

THEOREM 9.1. *Given $P \in \mathscr{P}$,*

$$\inf_{\mu} \left\| \pi P - \int \pi Q_x \mu(dx) \right\| \le \sup_x \|\pi H_x - \pi Q_x\| \equiv D, \tag{9.3}$$

where the inf *ranges over all probabilities on* $\mathbf{X}_2$.

PROOF. From (9.2),

$$\begin{aligned}\inf_{\mu} \left\| \pi P - \int \pi Q_x \mu(dx) \right\| &= \inf_{\mu} \left\| \int \pi H_x P(dx) - \int \pi Q_x \mu(dx) \right\| \\ &\le \left\| \int \pi H_x P(dx) - \int \pi Q_x P(dx) \right\| \le \int \|\pi H_x - \pi Q_x\| P(dx) \\ &\le \sup_x \|\pi H_x - \pi Q_x\| = D. \qquad \square\end{aligned}$$

The upper bound D in (9.3) should be quite good (as a universal bound) because $H_x \in \mathscr{P}$ for each $x \in \mathbf{X}_2$. Thus, if πQ_x is a reasonable approximation to πH_x, then D should provide a reasonable bound in (9.3). Here is the example of Section 8.2 reworked in the above notation.

EXAMPLE 9.1. Take $\mathbf{X}_2 = R^n$ and let $G = O_n$. Then, for $x \in R^n$,

$$H_x = \mathscr{L}(Ux),$$

where U is uniform on O_n. Since

$$Ux = \|x\| U\left(\frac{x}{\|x\|}\right),$$

H_x is the uniform distribution on

$$\{y \mid y \in R^n, \|y\| = \|x\|\}.$$

An easy calculation shows that

$$\mathbf{E}Ux = 0$$

and

$$\operatorname{Cov}(Ux) = n^{-1}\|x\|^2 I_n.$$

For this example, take

$$Q_x = N(0, n^{-1}\|x\|^2 I_n),$$

which is the normal distribution on R^n with the same mean and covariance matrix as H_x.

Now, let $\mathbf{X}_1 = R^k$ with $k < n$ and consider

$$\pi = (I_k \quad 0) \colon k \times n$$

as the projection from R^n to R^k. Given an O_n-invariant probability P on R^n, its projection on R^k is πP. According to Theorem 9.1, the variation distance between πP and the closest average of the πQ_x's is bounded above by

$$D = \sup_x \|\pi H_x - \pi Q_x\|.$$

But

$$\pi H_x = \mathscr{L}(\pi U x) = \mathscr{L}(\|x\| V),$$

where V is the vector in R^k which is the first k coordinates of a random vector in R^n which is uniform on the n sphere. Also,

$$\pi Q_x = N(0, n^{-1}\|x\|^2 I_k)$$

because

$$\pi\pi' = I_k.$$

Thus,

$$\begin{aligned} D &= \sup_x \left\| \mathscr{L}(\|x\| V) - N(0, n^{-1}\|x\|^2 I_k) \right\| \\ &= \left\| \mathscr{L}(V) - N(0, n^{-1} I_k) \right\| \\ &= \left\| \mathscr{L}(\sqrt{n}\, V) - N(0, I_k) \right\|. \end{aligned}$$

Therefore the calculation of D in this example reduces to finding the variation distance between the $N(0, I_k)$ distribution and $\mathscr{L}(\sqrt{n}\, V)$. An upper bound on this distance was given in Proposition 7.6 (with p replaced by k). This completes the example. □

Here are a few remarks concerning the above example which are also valid in other examples. First, Q_x is a terrible approximation to H_x; they are in fact mutually singular. However, πQ_x is a good approximation to πH_x. This is what matters in applications. Second, both Q_x and H_x are invariant functions of x so that the averages

$$\int Q_x \mu(dx), \qquad \int H_x \mu(dx)$$

can be written as averages over a maximal invariant under the group action on $\mathbf{X}_2$. In Example 9.1, a maximal invariant is $x \to \|x\|$ and the above averages are obviously just averages over $\|x\|$. Further,

$$x \to \|\pi H_x - \pi Q_x\|$$

is also an invariant function of x. This often makes the calculation of D (or bounding D above) an achievable task. In Example 9.1, the sup was in fact calculated explicitly by observing that for all $x \neq 0$,

$$\|\pi H_x - \pi Q_x\| = \|\pi H x_0 - \pi Q x_0\|,$$

where x_0 is a fixed nonzero vector.

The method described above is from Diaconis, Eaton and Lauritzen (1987) where it is applied to a variety of univariate and multivariate examples. In the examples discussed thus far, it is clear that the "appropriate" Qx's come from an associated "infinite" theorem. This is not at all clear in the description of the above method, but in every example that I know, there is some "infinite" theorem lurking in the background.

9.2. The general univariate linear model. The goal of this section is to discover the implications of extendability and invariance in the context of linear models. In a finite dimensional inner product space $(V, (\cdot,\cdot))$ consider an observation vector Y whose mean vector μ lies in a known linear subspace $M \subseteq V$ and whose covariance is $\sigma^2 I$ where I is the identity linear transformation on V. When Y is $N(\mu, \sigma^2 I)$, it is clear that

$$\mathscr{L}(Y) = \mathscr{L}(gY) \tag{9.4}$$

for all orthogonal transformations g such that

$$gx = x \quad \text{for all } x \in M.$$

The group of all such orthogonal transformations is denoted by $O(M)$. Rather than assume Y has a normal distribution, it is only assumed that (9.4) holds, namely, that the distribution of Y is $O(M)$-invariant.

Now, let $(V_1, (\cdot,\cdot)_1)$ be another finite dimensional inner product space and assume that π: $V \to V_1$ is a linear transformation which satisfies

$$\pi\pi' = I_1,$$

where I_1 is the identity on V_1. This π is an example of what we have been calling a "projection." With $Y_1 = \pi Y$, the mean vector of Y_1 is $\mu_1 = \pi\mu$ which is an element of $M_1 = \pi(M)$. Obviously M_1 is a linear subspace of V_1.

Given that $P = \mathscr{L}(Y)$ satisfies

$$gP = P, \qquad g \in O(M), \tag{9.5}$$

the problem we discuss here is: What can we say about $\mathscr{L}(Y_1) = \pi P$? The example discussed in Section 8.4 is a special case of this problem. In essence, the result below provides a bound on the variation distance between πP and a mixture of the distributions $\{N(\mu_1, \sigma^2 I_1);\ \mu_1 \in M_1,\ \sigma^2 \geq 0\}$.

The approach to this problem is that described in the previous section. To this end, let U have a uniform distribution on $O(M)$ and for each $x \in V$, write $x = x_1 + x_2$ with $x_1 \in M$ and $x_2 \in M^\perp$ ($M^\perp$ is the orthogonal complement of M). Then

$$H_x = \mathscr{L}(Ux) = \mathscr{L}(x_1 + Ux_2)$$

since $U \in O(M)$. It can be shown [see Diaconis, Eaton and Lauritzen (1987)] that

$$\mathbf{E}Ux = x_1$$

and

$$\operatorname{Cov}(Ux) = \frac{\|x_2\|^2}{n-m} C,$$

where n is the dimension of V, m is the dimension of M and C is the orthogonal projection onto $M^\perp$. Therefore

$$\mathbf{E}\pi Ux = \pi x_1,$$

$$\operatorname{Cov}(\pi Ux) = \frac{\|x_2\|^2}{n-m} \pi C \pi'. \tag{9.6}$$

To apply Theorem 9.1, pick Q_x to be the $N(x_1, \|x_2\|^2(n-m)^{-1} I_2)$ distribution on V. Then

$$\pi Q_x = N\left(\pi x_1, \|x_2\|^2 (n-m)^{-1} I_1\right)$$

is a normal distribution on V. Therefore

$$\begin{aligned} D &= \sup_x \|\pi H_x - \pi Q_x\| \\ &= \sup_x \left\| \mathscr{L}(\pi x_1 + \pi U x_2) - N\left(\pi x_1, \|x_2\|^2 (n-m)^{-1} I_1\right)\right\| \\ &= \sup_x \left\| \mathscr{L}\left(\pi U \frac{x_2}{\|x_2\|}\right) - N\left(0, (n-m)^{-1} I_1\right)\right\| \\ &= \left\| \mathscr{L}(\pi U x_0) - N\left(0, (n-m)^{-1} I_1\right)\right\|, \end{aligned} \tag{9.7}$$

where x_0 is any fixed vector of length 1 in $M^\perp$.

LEMMA 9.1. *Let k be the dimension of V_1 and set*

$$A_0 = \pi C \pi'. \tag{9.8}$$

For $k \leq n - m - 4$,

$$\left\| \mathscr{L}(\pi U x_0) - N\left(0, (n-m)^{-1} A_0\right)\right\| \leq 2 \frac{k+3}{n-m-k-3}. \tag{9.9}$$

PROOF. See Diaconis, Eaton and Lauritzen (1987), Proposition A.1. □

We now assume $k \le n - m - 4$ and A_0 in (9.8) has rank k.

THEOREM 9.2. *The variation distance between πP and the closest mixture of the normal distributions $N(\mu_1, \sigma^2 I_1)$ with $\mu_1 \in M_1$ and $\sigma^2 \ge 0$ is bounded above by*

$$\beta_n(A_0) = 2\frac{k+3}{n-m-k-3} + 2\left[(\det A_0)^{-1/2} - 1\right]. \tag{9.10}$$

PROOF. It suffices to show that D in (9.7) is bounded above by $\beta_n(A_0)$. But

$$\begin{aligned} D &= \left\| \mathscr{L}(\pi U x_0) - N\left(0, (n-m)^{-1} I_1\right) \right\| \\ &\le \left\| \mathscr{L}(\pi U x_0) - N\left(0, (n-m)^{-1} A_0\right) \right\| \\ &\quad + \left\| N\left(0, (n-m)^{-1} A_0\right) - N\left(0, (n-m)^{-1} I_1\right) \right\|. \end{aligned}$$

The first of the two summands is bounded above by $2(k+3)/(n-m-k-3)^{-1}$ according to Lemma 9.1. Because A_0 has all its eigenvalues in $(0, 1]$, it follows easily that

$$\begin{aligned} &\left\| N\left(0, (n-m)^{-1} A_0\right) - N\left(0, (n-m)^{-1} I_1\right) \right\| \\ &\quad = \left\| N(0, A_0) - N(0, I_1) \right\| \le 2\left[(\det A_0)^{-1/2} - 1\right]. \end{aligned}$$

This completes the proof. □

When $\beta_n(A_0)$ is small, Theorem 9.2 implies that πP is close to a distribution generated by first selecting (μ_1, σ^2) according to some distribution and then selecting Y_1 from a $N(\mu_1, \sigma^2 I_1)$ distribution. In other words, the smallness of $\beta_n(A_0)$ implies that Y_1 looks like it was drawn from a normal distribution. Thus the original invariance assumptions on Y have very strong implications for $\mathscr{L}(Y_1)$. These issues are discussed more fully in the next section where the standard univariate regression model is treated.

9.3. The regression model. In R^n consider the usual regression model

$$Y = X\beta + \varepsilon, \tag{9.11}$$

where X is a known $n \times q$ matrix of rank q, β is a $q \times 1$ vector of unknown parameters and ε is the error vector. Thus the regression subspace

$$M = \{\mu | \mu = X\beta,\ \beta \in R^q\}$$

is of dimension q. It is assumed that

$$\mathscr{L}(Y) = \mathscr{L}(gY) \tag{9.12}$$

for all $g \in O(M)$. This is equivalent to the assumption that

$$\mathscr{L}(\varepsilon) = \mathscr{L}(g\varepsilon)$$

for all $g \in O(M)$.

In this example, the "projection" π is taken to be

$$\pi = (I_k \quad 0): k \times n,$$

where I_k is the $k \times k$ identity and $k > q$. Partition Y as

$$Y = \begin{pmatrix} Y_1 \\ Y_2 \end{pmatrix}$$

with $Y_1 = \pi Y$. Also partition X as

$$X = \begin{pmatrix} X_1 \\ X_2 \end{pmatrix},$$

where X_1 is $k \times q$ so X_2 is $(n - k) \times q$. Finally partition ε as

$$\varepsilon = \begin{pmatrix} \varepsilon_1 \\ \varepsilon_2 \end{pmatrix}$$

with $\varepsilon_1 = \pi\varepsilon$ in R^k. Then the projected regression model is

$$\pi Y = Y_1 = X_1\beta + \varepsilon_1. \tag{9.13}$$

The statistical interpretation of this model description is the following. We do an experiment in which model (9.13) is assumed for Y_1. But we imagine that a larger experiment could have been performed and the invariance assumption (9.12) for the model (9.11) is assumed to hold. Then, the implications of this model assumption [namely (9.11) and (9.12)] are of concern. This is what Theorem 9.2 yields.

We now turn to the evaluation of $\beta_n(A_0)$ in (9.10) for the above regression model. First, the orthogonal projection onto the orthogonal complement of M is

$$C = I_n - X(X'X)^{-1}X'$$

so that

$$A_0 = \pi C \pi' = I_k - X_1(X'X)^{-1}X_1'.$$

Since the dimension of M is q in this example,

$$\beta_n(A_0) = 2\frac{k+3}{n-q-k-3} + \left[(\det A_0)^{-1/2} - 1\right].$$

Now, fix k and q, and let $n \to \infty$. In order to obtain an "infinite" theorem [i.e., $\beta_n(A_0) \to 0$ as $n \to \infty$] for this example, X: $n \times q$ must satisfy

$$\lim_{n\to\infty} \det\left(I_k - X_1(X'X)^{-1}X_1'\right) = 1. \tag{9.14}$$

Recall that k and q are fixed so we are thinking of X_1: $k \times q$ as a fixed matrix, namely, the design matrix of the experiment actually performed. With X_1 fixed, a necessary and sufficient condition for (9.14) to hold is that

$$\lim_{n\to\infty} (X'X)^{-1} = 0. \tag{9.15}$$

Equation (9.15) means that each element of the $q \times q$ matrix $(X'X)^{-1}$ converges to 0 as $n \to \infty$. The statistical interpretation of (9.15) is that the parameter vector β is consistently estimated (by least squares) since the covariance matrix of the least squares estimator of β is $\sigma^2(X'X)^{-1}$. This assumes that Y and hence ε has a covariance matrix which is $\sigma^2 I_n$ when (9.11) holds.

The point of the above discussion is that the conditions for the existence of an "infinite" theorem have a direct statistical interpretation in terms of the estimation of β. For a further discussion of the relationship between "infinite" theorems and statistical interpretations, see Lauritzen (1988).

Bibliography

GENERAL REFERENCES (BOOKS)

ANDERSON, T. W. (1984). *An Introduction to Multivariate Statistical Analysis*, 2nd ed. Wiley, New York.

BERGER, J. O. (1985). *Statistical Decision Theory and Bayesian Analysis*, 2nd ed. Springer, New York.

BLACKWELL, D. and GIRSHICK, M. A. (1954). *Theory of Games and Statistical Decisions*. Wiley, New York.

EATON, M. L. (1983). *Multivariate Statistics: A Vector Space Approach*. Wiley, New York.

FARRELL, R. H. (1976). *Techniques of Multivariate Calculation*. *Lecture Notes in Math.* **520**. Springer, New York.

FARRELL, R. H. (1985). *Multivariate Calculation: Use of the Continuous Groups*. Springer, New York.

FRASER, D. A. S. (1968). *The Structure of Inference*. Wiley, New York.

FRASER, D. A. S. (1979). *Inference and Linear Models*. McGraw-Hill, New York.

GIRI, N. (1975). *Invariance and Minimax Statistical Tests*. Hindustan Publishing Corporation, Delhi.

LAURITZEN, S. L. (1988). *Extremal Families and Systems of Sufficient Statistics*. *Lecture Notes in Statist.* **49**. Springer, New York.

LEHMANN, E. L. (1983). *Theory of Point Estimation*. Wiley, New York.

LEHMANN, E. L. (1986). *Testing Statistical Hypotheses*, 2nd ed. Wiley, New York.

MUIRHEAD, R. J. (1982). *Aspects of Multivariate Statistical Theory*. Wiley, New York.

NACHBIN, L. (1965). *The Haar Integral*. Van Nostrand, Princeton, N.J.

SEGAL, I. E. and KUNZE, R. A. (1978). *Integrals and Operators*, 2nd rev. and enlarged ed. Springer, New York.

CITED REFERENCES

ANDERSON, T. W. (1958). *An Introduction to Multivariate Statistical Analysis*. Wiley, New York.

ANDERSON, T. W. (1984). *An Introduction to Multivariate Statistical Analysis*, 2nd ed. Wiley, New York.

ANDERSSON, S. (1982). Distributions of maximal invariants using quotient measures. *Ann. Statist.* **10** 955–961.

ANDREWS, D. F. and MALLOWS, C. L. (1974). Scale mixtures of normal distributions. *J. Roy. Statist. Soc. Ser. B* **36** 99–102.

BERGER, J. O. (1985). *Statistical Decision Theory and Bayesian Analysis*, 2nd ed. Springer, New York.

BONDAR, J. and MILNES, P. (1981). Amenability: A survey for statistical applications of Hunt–Stein and related conditions on groups. *Z. Wahrsch. verw. Gebiete* **57** 103–128.

BOURBAKI, N. (1963). *Éléments de Mathématique Intégration*, Chaps. 7 and 8. Hermann, Paris.

DAS GUPTA, S. (1979). A note on ancillarity and independence via measure-preserving transformations. *Sankhyā Ser. A* **41** 117–123.

DAWID, A. P. (1977). Spherical matrix distributions and a multivariate model. *J. Roy. Statist. Soc. Ser. B* **39** 254–261.

DAWID, A. P. (1981). Some matrix-variate distribution theory: Notational considerations and a Bayesian application. *Biometrika* **68** 265–274.

DE FINETTI, B. (1931). Funzioni caratteristical di un fenomeno aleatorio. *Atti Accad. Naz. Lincei Mem. Cl. Sci. Fis. Mat. Natur.* (6) **4** 251–299.

DIACONIS, P., EATON, M. and LAURITZEN, S. (1987). Applications of finite de Finetti style theorems to linear models. Technical Report No. 502, School of Statistics, Univ. Minnesota.

DIACONIS, P. and FREEDMAN, D. (1980). Finite exchangeable sequences. *Ann. Probab.* **8** 745–764.

DIACONIS, P. and FREEDMAN, D. (1987). A dozen de Finetti-style results in search of a theory. *Ann. Inst. H. Poincaré. Probab. Statist.* **23** 397–423.

DICKEY, J. M. (1967). Matricvariate generalizations of the multivariate t distribution and the inverted multivariate t distribution. *Ann. Math. Statist.* **38** 511–518.

DUBINS, L. and FREEDMAN, D. (1979). Exchangeable processes need not be mixtures of independent identically distributed random variables. *Z. Wahrsch. verw. Gebiete* **48** 115–132.

EATON, M. L. (1970). Some problems in covariance estimation (preliminary report). Technical Report No. 49, Department of Statistics, Stanford Univ.

EATON, M. L. (1981). On the projections of isotropic distributions. *Ann. Statist.* **9** 391–400.

EATON, M. L. (1983). *Multivariate Statistics: A Vector Space Approach.* Wiley, New York.

EATON, M. L. (1985). On some multivariate distributions and random orthogonal matrices. In *Proc. Berkeley Conf. in Honor of Jerzy Neyman and Jack Kiefer* (L. M. Le Cam and R. A. Olshen, eds.) **2** 681–691. Wadsworth, Monterey, Calif.

EATON, M. L. and KARIYA, T. (1984). A condition for null robustness. *J. Multivariate Anal.* **14** 155–168.

EATON, M. L. and OLKIN, I. (1987). Best equivariant estimators of a Cholesky decomposition. *Ann. Statist.* **15** 1639–1650.

FARRELL, R. H. (1962). Representations of invariant measures. *Illinois J. Math.* **6** 447–467.

FARRELL, R. H. (1985). *Multivariate Calculation: Use of the Continuous Groups.* Springer, New York.

HALL, W. J., WIJSMAN, R. A. and GHOSH, J. K. (1965). The relationship between sufficiency and invariance with applications in sequential analysis. *Ann. Math. Statist.* **36** 575–614.

HERZ, C. S. (1955). Bessel functions of matrix argument. *Ann. of Math.* (2) **61** 474–523.

HEWITT, E. and SAVAGE, L. J. (1955). Symmetric measures on Cartesian products. *Trans. Amer. Math. Soc.* **80** 470–501.

HILL, B. M. (1969). Foundations for the theory of least squares. *J. Roy. Statist. Soc. Ser. B* **31** 89–97.

JAMES, A. T. (1954). Normal multivariate analysis and the orthogonal group. *Ann. Math. Statist.* **25** 40–75.

JAMES, W. and STEIN, C. (1960). Estimation with quadratic loss. *Proc. Fourth Berkeley Symp. Math. Statist. Probab.* **1** 361–380. Univ. California Press.

KARIYA, T. (1981). A robustness property of Hotelling's T^2-test. *Ann. Statist.* **9** 211–214.

KARIYA, T. (1984). An invariance approach to estimation in a curved model. Technical Report No. 88, Hitotsubashi Univ.

KARIYA, T. and SINHA, B. K. (1988). *Robustness of Statistical Tests.* Academic, New York.

KHATRI, C. G. (1970). A note on Mitra's paper "A density free approach to the matrix variate beta distribution." *Sankhyā Ser. A* **32** 311–318.

KIEFER, J. (1957). Invariance, minimax sequential estimation, and continuous time processes. *Ann. Math. Statist.* **28** 573–601.

LAURITZEN, S. L. (1988). *Extremal Families and Systems of Sufficient Statistics. Lecture Notes in Statist.* **49**. Springer, New York.

LEHMANN, E. L. (1959). *Testing Statistical Hypotheses.* Wiley, New York.

LEHMANN, E. L. (1986). *Testing Statistical Hypotheses*, 2nd ed. Wiley, New York.

MITRA, S. K. (1970). A density free approach to the matrix variate beta distribution. *Sankhyā Ser. A* **32** 81–88.

MUIRHEAD, R. J. (1982). *Aspects of Multivariate Statistical Theory.* Wiley, New York.

NACHBIN, L. (1965). *The Haar Integral.* Van Nostrand, Princeton, N.J.

OLKIN, I. and RUBIN, H. (1964). Multivariate beta distributions and independence properties of the Wishart distribution. *Ann. Math. Statist.* **35** 261–269. Correction **37** (1966) 297.

OLKIN, I. and SELLIAH, J. B. (1977). Estimating covariances in a multivariate normal distribution. In *Statistical Decision Theory and Related Topics II* (S. S. Gupta and D. S. Moore, eds.) 313–326. Academic, New York.

SCHWARTZ, R. E. (1966). Properties of invariant multivariate tests. Ph.D. dissertation, Cornell Univ.

SEGAL, I. E. and KUNZE, R. A. (1978). *Integrals and Operators*, 2nd rev. and enlarged ed. Springer, New York.

SELLIAH, J. B. (1964). Estimation and testing problems in a Wishart distribution. Technical Report No. 10, Department of Statistics, Stanford Univ.

SMITH, A. F. M. (1981). On random sequences with centred spherical symmetry. *J. Roy. Statist. Soc. Ser. B* **43** 208–209.

STEIN, C. (1956). Some problems in multivariate analysis, Part I. Technical Report No. 6, Department of Statistics, Stanford Univ.

TAKEMURA, A. (1984). *Zonal Polynomials.* IMS, Hayward, Calif.

WIJSMAN, R. A. (1957). Random orthogonal transformations and their use in some classical distribution problems in multivariate analysis. *Ann. Math. Statist.* **28** 415–423.

WIJSMAN, R. A. (1967). Cross-sections of orbits and their applications to densities of maximal invariants. *Proc. Fifth Berkeley Symp. Math. Statist. Probab.* **1** 389–400. Univ. California Press.

WIJSMAN, R. A. (1985). Proper action in steps, with application to density ratios of maximal invariants. *Ann. Statist.* **13** 395–402.

WIJSMAN, R. A. (1986). Global cross sections as a tool for factorization of measures and distribution of maximal invariants. *Sankhyā Ser. A* **48** 1–42.

ZIDEK, J. (1969). A representation of Bayes invariant procedures in terms of Haar measure. *Ann. Inst. Statist. Math.* **21** 291–308.